Thomas J. Burrill

The Bacteria

an account of their nature and effects, together with a systematic

description of the species

Thomas J. Burrill

The Bacteria
an account of their nature and effects, together with a systematic description of the species

ISBN/EAN: 9783337852894

Printed in Europe, USA, Canada, Australia, Japan

Cover: Foto ©berggeist007 / pixelio.de

More available books at **www.hansebooks.com**

THE BACTERIA:

AN ACCOUNT OF THEIR NATURE AND EFFECTS, TOGETHER WITH A SYSTEMATIC DESCRIPTION OF THE SPECIES.

BY T. J. BURRILL, PH. D.,

Professor of Botany and Horticulture, Illinois Industrial University; Botanist Illinois State Board of Agriculture.

CONTENTS.

INTRODUCTION.

INTRODUCTION.

Biology—the science of life—has no more interesting and important problems than those concerning the smallest and simplest of animate objects, the bacteria. These exceedingly common living things, visible to us only by the aid of powerful magnifiers, have latterly received much attention on account of the endeavor to find out by scientific research, the origin of life on our globe; but more for the undreamed of power and influence which they have been found to possess and exercise in the operations of the organic world. Since the discovery of their existence and modes of action, many questions previously unanswerable have become easy, and the knowledge obtained has passed into science to serve, not only as an intellectual stimulus to man, but as sure standing-ground from which he may reach other heights and gain at last his rightful dominion over the forces and objects of earth.

It is not too much to say that mankind could not continue to exist, could never have existed, on the earth as at present constituted, but for these minute though effective agents: the truth of this does not depend on any mere figure of speech, or even on any delicate adjustment of nature, but on the actual and essential work accomplished. Without them, and with other things as they are, there could have been no fertile soil, no luxuriant pastures, no bountiful harvests, no possibility for man. On the other hand we should not suffer many of the ills that flesh is heir to, if we could escape their restless activities in certain forms and ways.

It is the object of this paper to present, in language freed as far as possible from technical terms, the principal and most interesting facts now known about these silent-working denizens of the earth, the air, and the water. Should any one wish to inquire further, he may soon find an abundant literature on the subject; but the whole matter is still so new that a great proportion of it is only to be found in the scientific periodicals of the last two decades, and unfortunately for many, mostly in foreign tongues. The best general work of American publication is *The Bacteria*, a translation from the French of Dr. Magnin, by Dr. G. M. Sternberg: Little, Brown & Co., Boston, Mass., 1880. There are several excellent general articles in numbers of the Popular Science Monthly, and very instructive reports by Drs. Detmers, Law and Salmon, on infectious diseases of animals, in the annual reports of the United States Commissioner of Agriculture, for 1878, 1879, and 1880. For "blight" in pear trees, etc., see report for 1880, of the Illinois Industrial University; proceedings American Association for the Advancement of Science, xxix, (1880) p. 583. For the possible explanation of the origin of certain diseases, see "some vegetable poisons," (Rhus, etc.) l. c. xxxi (1882.)

Charles S. Dolley, of Rochester, N. Y., has issued in pamphlet form a translation from the German of an interesting general article by Ferdinand Cohn.

For the practical study of these minute organisms, a first-class compound microscope is required and the ability to use it; but any one possessing ordinary capacity for scientific study, with plenty of patience, may gain a large amount of information by the aid of an instrument costing not more than fifty to one hundred dollars.

Certain kinds of the organisms may be readily obtained by putting a' few bits of flesh or of any soft animal or vegetable matter in water kept at the ordinary temperature of a comfortable living-room and allowed to stand one to several days. Characteristic species often occur in given infusions. Others may be sought for in decaying fruits and vegetables, on the surfaces of cooked articles of food, as of boiled potatoes, rice, beans, etc., while some are commonly present in the mouth, especially in the collections on the teeth and the fur of the tongue. The pus from external wounds, and very often the discharges from pimples, boils, ulcers, etc., contain myriads of the organisms.

Bacteria may be mounted for preservation on the usual thin cover glasses used by microscopists, and the covers inverted and fastened to the ordinary microscopical slides. First smear the cover glass with the material containing the bacteria, and after drying a few minutes immerse in strong alcohol to fix the specimens by coagulation of the imbedding substance. Then stain by immersing in common violet ink (aniline) at least five minutes, wash in water and mount in a weak, aqueous solution of carbolic acid. There are many other processes and materials used to advantage in certain cases, but this is the simplest and most generally successful of any now known.

I have not considered it wise to encumber these pages with references, yet make no claims to anything original except as indicated in the proper places. I have, however, verified by experimental studies much of the information presented, gained in the first place from others.

PART I. NATURE AND ORGANIZATION.

1. EXISTENCE AND HABITAT.

The minute organisms, to which the term bacteria is now commonly applied, cannot be individually recognized without a magnifying glass of high power; so, to those not accustomed to work with the compound microscope, the existence of the little creatures as common objects may not be suspected, or at least rationally realized. To be sure much is said of the life of a drop of water or other fluid; but a large amount of this sort of retailed information is purely imaginary, having nothing or little of truth about it. The fact is, pure water, such as is taken from a good well or spring, has absolutely no life in itself, nor are there in it any living things whatever, no matter what magnifying power one examines it with. On the other hand, there are no other living things in nature, from the highest to the lowest, so widely distributed in the earth, the air and the water, throughout the world, so commonly and multitudinously present, near us, and on us and in us, as the various kinds of the microscopic organisms of which we now write. If pure well water does not contain these or any other living things, all foul waters do. Bacteria or their allies exist in countless myriads in all filth, in all decomposing vegetable or animal matter, whether ill-scented or not, in all organic substances undergoing apparently spontaneous change, as heating in the mass, becoming sour, rancid, or putrid, and, generally, undergoing those changes known as fermentation or decay.

The water from ill-scented cisterns, and indeed from many in which odor is imperceptible, contains varying numbers of these minute living creatures. The water of foul ditches, stagnant ponds, marshes and sloughs teems with them, in numbers surpassing those of the leaves of the forest; a drop under the microscope often presents a maze of living forms, more wonderful than imagination ever pictured, or of which fiction ever dreamed. The waters of running streams are more or less inhabited by them, and the ocean itself is the special home of peculiar kinds.

The air may, like water, be absolutely pure, or clouded by innumerable numbers of the almost imponderable, but living, multiplying things. Wherever there is floating dust of organic origin, they may be said to be certainly present. The stilled atmosphere of close apartments, and of thickly built streets, especially over decomposing filth, is laden with various forms of bacteria and allied organisms; and these are unavoidably inhaled by ourselves with every breath inspired. But in the dry, open country, few of them can be found in the pure, fresh, invigorating air, and over wide desert regions and the tops of high mountains none whatever exist. With us the air in winter, especially when the earth is long covered with snow, becomes almost free, while in midsummer and autumn, especially after damp and close weather, their myriad numbers are wafted to and fro by every movement of the freighted atmosphere.

The slime on vessels of standing water, on the surface of vegetable or animal substances, like cold articles of food set away for some time, e.g. boiled potatoes, cooked meats, etc., is composed of these organisms and their products. Brewers' yeast, used in bread making, is not a *Bacterium*, but it is a related growth, and the so-called salt-rising, not unfrequently used in making bread, owes its energy to living bacteria. Sour milk always contains swarms of the microscopic, moving things, and during the formation vinegar from the juice of fruits, from solutions of whisky and sugar, etc., they may always be found in similar numbers.

They are invariably present in pus from open, suppurating wounds, in the discharge from boils and tumors on and in our own bodies, and in the ill-scented accumulations of the bodily excretions, as from the arm-pits and unwashed feet. If they have any connection with bad odor, no one will wonder why the breath sometimes smells bad if the material collecting on the teeth or the fur on the tongue can be seen under a good microscope. The cleanest human mouth can hardly be said to be ever perfectly free from these active organisms: while those to which the tooth brush or its equivalent is a stranger, are veritable culture boxes, or hot houses, richly supplied with rapidly growing and prodigiously multiplying forms and kinds. We swallow them with our food, and at least some kinds sometimes retain their activity in the stomach and intestinal tube. It now seems certain that the latter is always inhabited by special kinds which have to do with the activities there in operation. In health the blood is usually quite free from them, but in certain diseases this too, as it rapidly courses through the arteries and veins, sweeps along in the current myriads of the minute but living and developing, ever active things, inappropriately called "germs." We may well say in such cases that "the blood is out of order."

What perhaps more than all else gives vivid interest to our microscopical studies of this sort, is the fact that special kinds of these tiny living beings are found uniformly in connection with the severest diseases known to man, either of himself or of the plants and animals about him. We may name as examples "blight" in pear, apple and many plants; hog and chicken cholera, pleuropneumonia and splenic fever in cattle; malaria, typhoid fever, diphtheria and small-pox in man. What the connection is between organisms and the diseases, we leave for another part of this paper. Very recently it has been fully demonstrated that consumption, the most terrible scourge of the human race, invariably has its accompanying specific form in the diseased tissues and the exudations from them.

It will thus be seen that these microscopic organisms are by no means unusually or uncommonly present about, upon and within us. Taken as a classified group they are more widely dispersed and dwell more numberously everywhere, in everything, than all other living things put together. For the truth of this statement I need only to appeal to the observation of any one who, having a good compound microscope, will take the trouble to state the facts as seen. Still it would be unfair to leave the impression that any one specific kind is thus invariably near us, or that the most common, taken as a simple species, is as widely and thoroughly disseminated as might be inferred from the foregoing. The statements made have reference to the numerous species comprised in the entire group, not to certain ones. It is one thing to say that birds are very widely dispersed over the earth; it is quite another thing to state the facts as they are concerning the common domestic fowl or the snowy owl.

2. Color, Shape, Size.

As a rule bacteria are white, so that when numerous in water the fluid is opalescent or milky. They very frequently settle down, when the proper food supply is exhausted, to the bottom of the containing vessel, and in this case form a white slimy or flocculent mass. But of numerous species each has its peculiar color, as red, blue, yellow or green. On the moist surfaces of prepared articles of food, as of cooked potatoes, bread, cheese, rice, etc., spots of a red color much resembling drops or splashes of blood are not uncommon. Formerly these reputed blood-stains were regarded with superstitious wonder, and have been held to indicate guilt, and the anger of

God. These, as well as other colored spots and pigments, are now traced to their true cause, and are found due to the distinct color of the organisms themselves or to the chemical combination which they produce.

The different kinds of bacteria vary in shape from spherical to oval, cylindrical and thread-like; and the latter are straight, or crooked, or spiral, flexible or rigid. The spherical ones are often connected in two's or more, and sometimes form bead-like strings or chains. The main classifications in use are based on the form and size of the organisms.

It was impossible to know anything about bacteria until after the invention of the microscope, because none of them can be individually seen with the unaided eye. In transverse diameter, one twenty-five thousandth of an inch is a very common measurement, while some, including spherical ones, are even less than half this size. Now, a dot one two-hundredth of an inch across is barely visible to the eye of most persons, hence a magnifying power of more than one hundred times across (ten thousand in area) is required to barely see a common-sized *Bacterium*. To make out its real shape and any details of structure, ten times the enlargement mentioned is necessary, and not unfrequently as much more as can be secured by the highest possible powers of the microscope. Increase the heighth of an ordinary man one thousand times, and his head would be over a mile above the earth, yet under the same magnification one of these organisms would have plenty of room to swim freely, to stand on end and dance up and down, in the film of water included between two pieces of flat glass pressed so close together as to strongly adhere by capillary attraction. From one hundred to two hundred and fifty of them placed side by side would be required to stretch across the ordinary thickness of book paper. They are the smallest living organisms known to man, yet, as we shall see, by no means the least important in the economy of nature

3. Movements.

Bacteria have sometimes been divided into groups upon their apparent ability to move or not; but further study has demonstrated that many, if not most, species have states in which they remain at rest, and others in which they are freely motile. These states depend partially on their stage of development, partly on the surrounding conditions. For the latter the degree of moisture and temperature, and the food supply, are especially effective.

Some species at most only oscillate and quiver in the fluid medium in which they grow, never making progress in any given direction; others slowly and smoothly glide along in a straight but more often undulating path, while still others whirl and dance and roll, turning over and on end, now spinning round and round, now swaying gently back and forth, now darting like a flash across the microscopic field. Sometimes they move as though perfectly free, and had abundant muscular force, at other times they appear to be struggling to overcome obstructions, or to free themselves from some impediment. Not unfrequently they may be seen to carry

7

along little adhering extraneous particles, well showing their vital power, or two, in some way attached, pull in opposite directions with varying advantage for the one or the other.

No one, having once seen these motions, can doubt the inherent power the little things possess, or can question their right to be classed as living objects; whether as animals or plants, will depend on his previous information and experience, as well as upon what he sees. How they move, that is by what sort of organs or mechanism, is not easy to make out; but they have this power only when immersed in a fluid medium. When dry they are motionless, except as carried by external agents, as air currents.

It is now well known that different species have different methods and facilities for dissemination. Many kinds are externally viscid, or are always, when moist, imbedded in a mucilaginous exudation, and hence in either case adhere to any solid substance they may touch. These therefore are rarely found floating in the air. So long as the material containing them does not become dry and then pulverized, such species are not distributed by the currents of the atmosphere. It is quite possible that a foul drain or a filth-filled cesspool may contain myriads of disease-producing bacteria which are, however, only taken into the human body by swallowing them in contaminated water, the vitiated air though laden with fetid gases being harmless, yet this can by no means be said of all species known to be injurious to man. Their great numbers, their exceedingly minute size and their powdery character in mass, together in many cases with their long continued vitality, pre-eminently fit them for aerial distribution. The dust made evident by a sunbeam is often in considerable portion composed of these specks of living matter, which only await suitable conditions for growth and development to take place. What intellectual light the careful study of such a sunbeam may throw upon knotty questions of vital importance to man!

4. STRUCTURE.

Small as these organisms are, they possess well differentiated parts, which are each, presumably, absolutely necessary for their existence as living things. We do not infer that structure is life, but there is every reason to believe that life is just as dependent on structure in these simple and low forms as it is in the highest plants or animals. Solution, separation of parts, can no more take place in the one than in the other, without destruction of life. Liquids, separated from solids, never possess, nor are possessed by, the life principle, whatever that is. Think of dissolving the body of a mouse in acids, and then, by proper chemical manipulation, collecting every atom of the original substance and reforming by this means a living mouse! There is, however, just as much ground for the supposition that this can be done with the mouse, as there is in the case of the minute things of which we now write. To be sure, there is not so much complexity of organization, but there is quite as constant a certain and characteristic structure on which life depends.

Bacteria, like other living things, are composed of cells; or, perhaps, it is better to say, each *Bacterium* is an organic cell. There is an outer wall or membrane closed on every side like the skin of a raisin, and within this envelope there is a plastic substance forming the entire contents of the cavity. Chemical reagents show that the outer membrane is made up of *cellulose*, the substance of which wood consists, everywhere forming the framework of all plant structures. This, at first, is uniformly white, but becomes variously stained, as in the heart-wood of trees of different kinds. It is, except in peculiar states, insoluble in water, but swells and shrinks from its power of absorbing and again giving up this liquid. Fire destroys it, forming, with the oxygen of the air, carbonic acid and water.

The inner soft substance of the cell is *protoplasm*, which is even more common in higher organisms than cellulose; for while the latter is almost without exception confined to plants, the former is present in all living things, plants and animals alike. Taken by itself, it probably has a uniform composition in all organic nature; but it is variously mixed with other substances, rendering it difficult to determine the exact chemical components. Besides the carbon, oxygen and hydrogen of cellulose, it invariably has nitrogen, which cellulose does not have. Though usually semi-fluid, it also is insoluble in water. Heat coagulates, and thus destroys its functional activity—kills the organisms. Not the least instructive is the power of motion possessed by protoplasm. Its mass is often agitated by internal currents, made evident by little granules carried along by the stream. Sometimes these granules remain for a time at rest, then start off in an irregular line across the cell, rolling and tumbling on the way. These motions are not easy to see within most bacteria, but are readily enough made out in some of the larger kinds.

This is the simple structure of a *Bacterium*—a minute cell with an outer cellulose wall inclosing the protoplasmic contents. There are no organs or appendages, save in some, and perhaps all motile forms, there is at one, or at most, two points, an exceedingly fine, hair-like appendage, to the vibrations of which the movement of the organism is attributed. This fine *cilium* or *flagellum*, as it is called, is a difficult thing to see, even with the best microscopic equipment and most expert manipulation, partly on account of its exceeding fineness or thinness, partly from its rapid vibrations. Sometimes, by introducing a weak solution of iodine, so as to gradually subdue the movement, the cilium can be made out when not practicable otherwise.

5. Reproduction and Development.

Bacteria increase by one dividing itself into two. There is no such thing as sexual differentiation, nor is there in the development anything comparable to the germination of a seed. Division is affected by the formation of a transverse partition or septum of cellulose across the middle of an adult cell, thus making two compartments, after which separation gradually takes place by the parting of the outer wall and the middle of the new partition, with a rounding off of the contiguous ends; while, during the same time, each half grows to the size of the original whole. Thus one becomes two, each of the latter being in every way similar to the first. Under favorable circumstances this process may in some species take place within one hour, then in the course of another hour be repeated, and so indefinitely. Shorter periods for the process have often been reported. If we pause to rationally comprehend this rate of increase, we shall soon be lost in the amazing numbers which in a little time spring from one. It may seem incredible, but any one may readily verify the fact that according to the rate of one division each hour, the number from one after twenty-four hours will be more than sixteen millions, and after forty-eight hours nearly three hundred billions. Cohn has computed that *Bacterium termo* (a very common species in putrifying organic matter) would, at the end of twenty-four hours, at the above rate of increase, fill a little cube one-thousandth of an inch across each side; after forty-eight hours the solid mass would amount to about a pint; then of course at the end of another hour—since each would become two in this time—there would be two pints, in two hours four, in three eight, in four sixteen, etc., and such is the astonishing increase by this geometrical progression, that at the end of five days from the beginning of the mass arising from one of these exceedingly minute creatures would be sufficient to fill completely full, or to equal in weight the water of, the oceans of the world.

Incredible as this seems it is a simple calculation which any one can make on the supposition that the organism is one twenty-five thousandth of an inch in transverse diameter and twice as long, and that the number is double every hour. That nothing like this ever really occurs in nature for any consecutive number of hours, we need not be told; but the want of it only shows the harmonious interaction of causes and effects among the sum total of forces and activities on our earth; it is not because this *Bacterium* under the assumed conditions does not increase at the marvelous rate stated, but because such conditions do not continuously exist. The limitations are fixed by the given supply of proper nutriment, the absence of harmful chemical compounds, temperature, electrical and mechanical agitations, the effects of other living organisms, etc.; nevertheless the computations prepare us for the acceptance of facts, which would otherwise be deemed incredible, and they aid us in the explanations of observed phenomena otherwise inexplicable. When it is known that a certain species may reproduce itself with anything like this rapidity, we need not wonder that individuals are found, after a short exposure of the nutriment substance, in innumerably great numbers; and, if there is any chance for one or more of such individuals to gain entrance in the first place, we need not resort to speculations concerning spontaneous generation, or the transformation of parts of other living things to account for the multitudes subsequently found.

But some species seem never to grow or multiply rapidly, even under the most favorable conditions. They differ among themselves as do the species of the higher plants. Some varieties, for instance, of the cultivated radish may have two or even three generations in a season; while others require two years to perfect seed. Some weeds come to maturity in a few weeks and produce thousands of seeds, while an oak lives through the centuries.

In the process of self-division, complete separation may or may not take place. There is indeed considerable variation in this respect, even in the same species, according to the external and perhaps internal conditions. If, after the transverse dividing septum has been formed, further separation does not take place, the result is in a little time a filament, made up of cells placed end to end, and either quite smooth and of uniform diameter, or more or less constricted at the joints, so as to be bead or chain-like. In this case, however, each link is really a distinct individual so far as its physiological functions are concerned, and life is in no way interfered with if they are mechanically or otherwise separated. In one genus (*Sarcina*) division appears to take place in two directions · transverse and longitudinal—so that little regular squares of cells are produced, four or sixteen or more being frequently seen associated together in order.

The individuals of certain species cohere together without order in a somewhat compact mass called *zooglœa*, bound by the glairy · substance in which they are imbedded. But this is the case only in special conditions or states, for it is easy to see single ones free themselves from the immovable mass, and swim away in the liquid in which the whole are submerged. The zooglœa mass may be the result of the multiplication of a single cell, or those originally separate may become thus united.

Besides the method by division, several species are known to reproduce themselves under certain conditions by what are termed *spores*. The protoplasm of a cell becomes consolidated into a spherule occupying only a small part of the cavity. This afterward becomes free by the dissolution of the old cell wall, and, at the proper time, under the proper conditions, the spherule develops into a full-sized cell like its parent form. The formation of such so-called spores usually takes place when the nutriment is nearly exhausted, or when not suited, either in kind or in the required state as to temperature, etc., to the regular growth of the organism. It is a method of self-perpetuation rather than multiplication. Still, in a few cases of which record has been made, more than one spore is produced by each cell. Touissant, of France, has seen within the cavity the formation of spore sacks (*sporangia*), each of which produced from three to six spores. It is often in the form of these all but infinitesimal spores that bacteria are carried in the atmosphere. When produced in liquid they fall to the bottom after the dissolution of the old cell wall, and form, when numerous enough, a plainly visible, white sediment. In case the water is then evaporated, the sedimentary material becomes a fine powder, the particles of which readily float in moving air.

6. VITALITY AND ENDURANCE.

Moisture is essential to the growth of bacteria, but prolonged drying does not necessarily kill them. In the case of the species causing the blight of pear and other plants, they certainly lose little or none of their vitality by preservation during two years in the dry state on the point of a quill. Among those kinds producing spores, the adult organisms often do not live more than a few days or weeks, while the spores seem to absolutely defy time. Pasteur

thinks he has shown that the latter retain their vitality in the earth
out of doors for at least twelve years. After keeping in a sealed
tube four years, some virus of splenic fever was sent from England
to this country, where it was opened and used with fatal effect in
inoculating animals. It has even been argued by some investigators that bacteria never die, except as destroyed by fire or injurious
chemical compounds; and whatever may be the case otherwise, so
far as age is concerned this is true, for on account of their manner
of propagation an individual does not grow old.

Each species is adapted to a certain heat temperature from which
variation may take place, within only very narrow limits in some
kinds, or a wider range with others, (75° to 105° F.) None appear
to be killed by cold, but their vital activities cease at given limits,
never reaching below the freezing point of water. An artificial temperature of 123° F. below zero has been tried without killing them.
On the other hand not only is the limit of activity reached by increase of temperature, but death always ensues by exposure to a
given but greatly varying degree of heat for different species. Spores
immersed in the fluid resist for a time the heat of boiling water, but
all adult forms are killed in this way. When the medium is alkaline a higher degree of heat is required to destroy life than when
neutral or acid, and if the organisms are dry a still higher temperature may be withstood; 140° F. in water, fruit juices, vinegar, sour
milk, etc., is absolutely fatal to such as ordinarily occur in these
fluids. The highest temperature yet known at which any bacteria
are able to live and develop is 133° F.; this has been observed in
an infusion of beans. But some spores are not killed, especially in
alkaline fluids, by prolonged boiling, just as some seeds withstand
a similar test. If either are softened by germination, they quickly
succumb.

We make use of these facts in the operations of every day life,
especially in the process of canning articles of animal and vegetable
foods, scalding cooking utensils, etc. The heat kills the bacteria
present.

Various substances destroy bacteria, or at least prevent their
propagation in nutritious solutions. Of these, the best known and
most widely used are carbolic acid and quinine. Several others are
more deadly to the organisms than these; but they either have some
objectionable properties, or have not been so generally introduced.
Grace Calvert asserts that some bacteria can live in pure carbolic
acid, but there seems to be some mistake in this, at least for ordinary species, for hundreds of experiments prove that none are developed in a liquid containing one-half per cent. of this antiseptic.

The following table, from experiments by M. Jalan de la Croix
(Revue Scientifique, Feb. 4, 1882), shows the number of parts of
water to one of the substance named which barely permits the
development of bacteria in infusions of meat. For instance, if thirty
parts of water are mixed with one of alcohol, organisms may infest
the mixture; but if twenty-nine parts of water are used to one of
alcohol, bacteria do not develop, and the mixture does not ferment:

Alcohol	30
Chloroform	134
Soda biborate	107

Eucalyptol	308
Phenol (carbolic acid)	1002
Thymol	2229
Potash permanganate	3011
Picric acid	3011
Borated soda salycilate	3377
Benzoic acid	4020
Etherial oil of mustard	5734
Sulphurous acid	7534
Alum acetate	7535
Salicyclic acid	7677
Mercury bichloride	8358
Lime hypochlorite	13092
Sulphuric acid	16782
Iodine	20020
Bromine	20875
Chlorine	34509

From my observations these results seem to be trustworthy, though others very dissimilar have been published. It is not a sufficient test to observe through the microscope the effect that any substance has upon the motions of bacteria. Sometimes the addition of pure water causes their motions to cease, and sometimes molecular oscillations, not always easy to distinguish from those of life, continue after the organisms are dead. This has no doubt been a source of error in some of the accounts given to the press. The above figures were arrived at, by trying more and more water to one gramme of the substance until at last bacteria developed and putrefaction set in.

It should be said that it is well nigh impossible to kill bacteria in the air, by any kind of fumigation, except in a very thoroughly closed space, like an air-tight vessel or box. A mistaken sense of security often exists when the atmosphere of living rooms, sick chambers, etc., are pervaded by the odors of some supposed antiseptic material. These floating "germs" can at least withstand as much fumigating as any human being, and usually very much more. On the other hand, much can be done towards destroying noxious bacteria or preventing their growth and propagation, by impregnating liquids and solids by proper amounts of such destructive agents as are enumerated above. Experiments are now in progress of treating the carcasses of slaughtered animals with antiseptics not poisonous to man, for the purpose of sending fresh meat without ice across the ocean or other similar distances, and the results give much promise of success. A white powder is offered for sale to keep fruit without canning, and it appears to have virtue enough to really accomplish the result claimed without detrimentally affecting the food. (Compounds of boracic acid). There is, no doubt, much to be gained of practical utility in this direction.

7. Nutrition.

Bacteria absorb their food by endosmosis through the cellulose coating of their minute bodies. The nutrient material must therefore be in a liquid state as well as of the kind and strength suited to the specific organism. Like more highly organized living things, each species has its own peculiar kind of food, without which development does not take place. As a cat would starve on hay, and an ox on meat, so similar differences are found among these inhabitants of the microscopic world. But all require organic food derived from the bodies of higher plants or animals, though it has been sufficiently proved that at least some kinds can take the required nitrogen from inorganic sources. Bacteria could not live alone on the earth; they are not destroyers, not up-builders; they pull down and decompose material prepared by the assimilation of other vitalized workers.

Some kinds are rigidly confined to dead matter, others develop and grow at the expense of living plants or animals, causing little or no inconvenience or even conferring real benefit to the latter; or instituting varying degrees of discomfort, injury and disease. The waste products of their physiological processes are varied, after their kinds, as their food,—acids, alkalies, gases, liquids fragrant, ill-scented, etc., but always of simpler chemical composition than the original, until at last, through perhaps the consecutive action of several species, the elemental state is reached. Thus in the fermentations of solutions of sugar, alcohol is given off as a waste product by the "yeast plant" (*Saccharomycetes*) and the diluted alcohol ferments through the effects of the "vinegar plant" (mycoderma), and in this case vinegar is the substance thrown off, while this in turn decomposes in a similar manner by other agents into carbonic acid and water.

8. Origin.

The question, "Where do bacteria come from?" that is, what is their original source, is sure to press itself for answer. This answer is by no means an easy one to give with postive assurance of correctness. We have already seen how they are reproduced and the rapidity with which they may multiply, as well as something of the modes by which they are disseminated; but all thinking persons wish to know how or whence the first ones came into existence.

Upon this question there have been many experimental inquiries by the ablest investigators, and many theories have been propounded by those most competent to formulate them, as well as the crudest of unsupported hypotheses by untrained minds. Among all these opinions we may select three as worthy of special notice:

1st. Spontaneous formation under proper conditions from inorganic chemical elements or compounds. 2d. Transformation from other living organisms or their parts; and 3d. Direct creation by the great Author of all things.

It would require too much space to even sketch in this place the arguments which have been advanced in support of each of these theories by thorough scientists. The voluminous literature on the subject shows the interest that has been felt and the real warfare of words that has been waged by contending supporters of their own or other ideas.

But no one without personal bias can carefully review this literature to-day, with the results of experiments clearly in mind, without concluding that the first and second have not been proved. From a philosophical stand-point the first is highly plausible, since these organisms are admittedly among the lowest living things. If spontaneous generation takes place anywhere in nature, we should look for the phenomenon here; but, while certain experimenters have supposed their results showed the theory true, others still more expert have uniformly pointed out the fallacies of such experiments or deductions, so that up to the present time we are bound to say the facts known disprove rather than prove the hypothesis. The second has not been so fully and conclusively studied, yet the most searching investigations have not been wanting, with, upon the whole, negative results. No untrammeled scientist conversant with the entire subject as now known, admits the proof of the origin of bacteria by transformation of the parts of other plants or animals, however close and constant may be the accompaniment of the one with the other. In this statement we exclude the idea of evolution during centuries of time, the change taking place by numerous unrecognizable differences; for, to our mind this belongs under the third hypothesis. We have, therefore, just the same scientific information about the origin of bacteria as we have about that of other living things, man included. Whether we hold that the Creator fashioned each species in the beginning by a direct act, or accomplished the same by instrumentalities, working through the ages, makes no difference in the question before us. We are forced to refer the origin of bacteria to the same Power to which we attribute the origin of man. It is certain that these minute organisms have existed in very closely similar, if not identical, forms as at present, from remote antiquity, as is proved by the discovery of fossil remains, as well as the evident effects produced in the past ages of the world, according perfectly with those now known to be results of these creatures.

It must be admitted, however, that in the progress of individual development some kinds pass through different forms, which have been supposed to be characteristic of species, and that, retaining the same form, the physiological effects may vary through the prolonged influence of certain external conditions. Thus an individual cell may be spherical, then oblong, cylindrical, and filamentous, in regular sequence of growth, though each of these forms in other species may be characteristic and unchanging. And a species ordinarily living in an infusion of dead vegetation may, by a slow and gradual change, become capable of surviving and multiplying in the blood of living animals, whereas a sudden transfer from the one to the other medium would have permitted no such results. But man himself is capable of becoming adapted to as remarkable changes. We say we become habituated or acclimatized to things and conditions, as of eating arsenic and of dwelling in the malarial regions of tropical countries. No greater changes than these are known in the life history of bacteria, only that it requires less time for the modifications to take place; and this should be anticipated from their rapid succession of generations.

9. Place in Nature.

Bacteria are certainly living things, hence, according to the ordinary thought and classification of objects, must be either plants or animals. Formerly spontaneous or self-caused motion was held to be characteristic of animals as distinguished from plants; and, as many bacteria are seen under the microscope to move freely and in some cases very rapidly to and fro, hither and yon, they were at one time unhesitatingly classed as animals. as indeed most unscientific observers would at once pronounce them now when viewing them for the first time by the aid of a powerful magnifier. But the fact is, self-caused motion is not confined to animals. Really all plants more or less possess this power. A seedling causes its stem to turn upward and rootlet downward through internal forces, and afterward the growing parts bend to or from the light, as every one has observed. Leaves quite generally assume different positions day and night; flowers open and close, twining plants revolve in such manner that the free end sometimes sweeps a circle of four feet in diameter about the supporting object, tendrils twine in some instances fast enough to be plainly observed by any one possessing good eyes and a fair share of patience. It is true that all the higher plants are fastened to their supporting substance, and are incapable of roaming from place to place; but the lowest members of the vegetable kingdom are not thus limited. In some conditions at least the latter move as freely as any animals, not simply through the controlling influence of external agents, but from forces wholly within themselves. Such motile plants are all microscopic in size, but no botanist or zoologist hesitates to pronounce them true plants, having all the fundamental characteristics of members of the vegetable kingdom.

So, freedom of motion itself, however striking it may be, is not sufficient cause for classing bacteria with animals. But it has been before stated that the outer coat or wall of the organisms is composed of cellulose, and this is peculiarly a substance belonging to all plants, and, with very rare exceptions, not to any animals. It has also been mentioned that bacteria are capable of taking the nitrogen required for their nutrition from inorganic salts, and this is alone characteristic of plants; there are no exceptions on the animal side.

These and other reasons have in recent times caused all naturalists who have made these objects a special study to pronounce them plants. There is no difference of opinion among proper authorities, if the choice is limited to one or the other of the two great kingdoms of animated nature; but Haeckel, a German scientist, has proposed to take the lowest forms of both these kingdoms and constitute therewith a third, called *Protista*; in this case the bacteria would be called neither plants nor animals.

As plants, the bacteria certainly occupy a position at or very near the bottom of the scale. They are among the simplest in structure of living things, and include among them the smallest objects in nature, animated with a spark of that vitality whose nature and essence is as unknowable in them as it is in the being created in the image and likeness of God.

There is a further and undecided question as to whether the bacteria belong to the *Fungi* or to the *Algæ*, two great divisions of the vegetable kingdom; but this depends solely on the definitions given to these groups. If greater prominence is given to form and habitat, they more nearly resemble the Algæ, or sea-weeds; if physiological functions are decisive, then the bacteria are unquestionably Fungi. This is now less important, since in a strictly natural classification it is undecided whether or not the great groups known as Algæ and Fungi are entitled to remain as such. Probably not; in which case the term *Protophytes* will almost surely be uniformly adopted to designate the lowest division of the vegetable kingdom, and in this the bacteria will presumably occupy the lowest place under the original name of *Schizophytes*, a word now in use, and meaning dividing or splitting plants.

—

PART II. EFFECTS OF BACTERIA.

1. FERMENTATION AND PUTREFACTION.

Until within recent times it has been supposed that organic matter, dispossessed of the vitality to which it owed its existence, was naturally very unstable in its chemical constitution, and that its tendency was to go back to the elementary inorganic state, through the operations of simple chemical affinities. The soft parts of animal bodies have been considered especially liable to change after death; indeed this is now so common and constant a phenomenon that it is looked for as a matter of course, unless express provision is made to prevent it. We are surprised when told that during the summer season the fresh flesh of buffalo on the western plains remains sweet and perfectly good until dry in the warm open air. The astonishing story is repeated again and again that the bodies of the woolly elephant of ancient Siberia have been taken out of the ice, in which they perished long ages ago, yet with flesh still so fresh that dogs feed upon the carcasses.

The scrupulous care which we are obliged to take to avoid ill consequences to butchers' meats, makes these exceptional instances of preservation in nature really wonderful to us. We say the law is for such material to putrify and decompose, for milk and cider to sour, for the expressed juices of fruits to ferment, and the fruits themselves to decay. A pile of green herbage heats and rots, and wood exposed to moisture gradually loses its strength and disappears to help form vegetable mold.

These various changes, to which all dead organic matter is subject, going on about us so abundantly and so constantly, make up a large part of the physical phenomena of the world; we expect them to occur and recur with the same certainty if not the same regularity of time or rate as the fall of unsupported bodies to the earth, the planetary changes and the succession of the seasons. We just as much expect fresh meat to spoil in warm summer weather as

we do enkindled wood to burn. The wine maker counts just as
certainly upon the fermentation of the juice of the grape as the
engineer upon the pressure of superheated steam, though neither
the one nor the other may stop to consider the philosophy of the
phenomena with which they are respectively confronted. They have
ascertained by experiment the governing conditions, and proceed with
the confident assurance of what has been, will be.

Now, to those who herein read for the first time that dead organic
matter has in itself no such tendency to spontaneous change, that,
subject, as in nature generally, to all the activities of pure air, and
pure water, and to these alone whatever the temperature, dead fish
and flesh will not become ill-scented or putrid, that milk and blood
will not change from the condition they have when drawn from the
living animal, that a heap of green or wet grass will not heat and
rot, that moist wood will remain as durable as granite, and that
the substances of their own bodies after life has departed are as in-
corruptible as gold,—these words may seem foolishness and upon
the face of them absurd; yet this is the teaching of science, and is
the unavoidable conclusion from many instructive experiments.
When the whole facts are known, the wonder is rather that the flesh
of slaughtered animals so surely putrifies with us, not that as a rule
fresh meat exposed to the pure air of the western plains should remain forever good. These facts could never have been known
without the aid of the microscope, and since this wonderful instrument is of modern invention, the knowledge set forth in the following pages has been gained alone by modern investigation. If there
is still doubt about the matter as a general phenomenon, it is only
because new ideas are slowly accepted; if there is dispute among
the informed as to details of the process, it is mostly because so
few really competent experimenters have yet undertaken the delicate
but fruitful work.

The marvelous progress of modern science is based on the well-
grounded idea that every effect has an adequate cause, and that
these causes, in the material world at least, are subject to unde-
viating law. If a body moves, the force is sought, and usually not
in vain, which produced the motion. If change occurs, a competent
agent is at once supposed to be instrumental in its accomplishment.
Students of nature are not content with passing anything as mysterious which can be brought within the domain of knowledge, nor
with accepting as a fact anything which does not fall within the reign
of natural cause. Possessed of this spirit, and provided with the
necessary instruments and means, the subject before us could not
escape investigation by the quickened intellects of recent times.
The result is, after much conflict of opinion and difference of inter-
pretation, the established fact, that *the natural changes taking place
in non-living organic matter, are all due to the vital activity of living
things.* Some of the usual results of life-forces may be accomplished in the chemist's laboratory, but the processes and conditions
there and in nature are entirely different. What life is, and to
what its particular powers are due, we do not know; but we do
know its effects, and these are as pronounced and unique in the
natural destruction as they are in the original upbuilding of organic
matter. Life manufactures, and life in turn pulls to pieces and
—2

destroys. An organic body is not a watch, which, having been
wound up, runs down of itself; but it is a splendid temple, the rich
material of which, accumulated from all lands, would require the
same as its original freighting for its redistribution. But the low,
microscopic organisms are by no means alone, if they act in any
sense different from other vitalized beings, in the work of destruction. Every living creature is continually destroying itself, reducing,
through its physiological and normal processes, the solid parts to
liquids and gases, from the organic to the inorganic. This is the
waste which all plants and animals suffer as long as life continues.
After death waste goes on in a different way, through the physiological and normal activities of other living beings, and the more
noticeably because there is no repair.

Among these latter destroyers there are very many kinds of
animals and plants. Indeed all animals are included in the list, and
the digestion of food with them is always a work of destruction, as
is readily understood. There are among flowering plants certain
kinds which are also purely destroyers—the dodder or golden thread
(Cuscuta), found sometimes in tangled profusion on weeds, flax, clo-
ver, etc., is common with us. But it is to the Fungi that we must
look for the principal agents, of the plant kind, which act as pure
destroyers of organic matter. These degraded plants live solely on
the accumulated and organized products of other plants and animals,
assimilating a portion for the architecture of their own bodily
structure, and exhaling another very considerable part as waste, in
one shape or another, but ultimately as carbonic acid and water,
two prominent ingredients in the original nutrition of green-leaved
plants. Nitrogenous compounds, as ammonia, nitric and butyric
acids, are also given off in the destruction of most organic matter.
An old log in the woods having no tendency to decay, and resisting
much better than iron the slow corrosion of the oxygen of the air,
tumbles to powder under the digestive power of insects, toadstools
and bacteria, each kind working differently, but accomplishing
nearly the same result.

It has already been said that in their physiological effects at
least the bacteria are Fungi. Their food is organic, elaborated in
the first instance, if not direct, by green-leaved plants. Their
function is to destroy, like that of other colorless plants and all
animals. In this process of destruction peculiar and characteristic
effects are usually produced by each species, and in very many
cases each species is limited to some special kind of food material.
In this there is nothing new or strange, for the law holds good
throughout all nature, among all animals and all plants.

If, now, we remember the facility of distribution which these
minute organisms enjoy, their vital endurance and their wonderful
powers of reproduction, we need not after all be surprised that
milk, wherever left exposed under the proper conditions of temperature, etc., becomes sour through the agency of a living organism
developing in, and feeding on, some element or elements of its
substance; or that fresh meat becomes ill-scented through the
respiration of similar living things, acting in a similar way. Keep
these destroyers out, and no such results will occur. In wide arid
regions where there is but little material on which they may

develop, their uniform presence in the air or on the surfaces of solids, cannot be expected, and this is the secret (now a secret no longer) of the fact that meat keeps in such places without putrefying. Since the white man has made his habitation in the West where the old hunter used to expose his jerked buffalo with impunity to the warm sun and air, this can no longer be done with butchered flesh. But it is still found possible on these wind-swept plains to keep meat fresh for a considerable time by sticking it on poles high above the earth, above the usual dissemination in these places of living organisms. The same thing is true on high mountains, though the warmth of the sun may be fully sufficient during mid-summer for the manifestation of all kinds of life and of rapid changes in putrefactive substances.

Among the destructive alterations of organic matter, those known as fermentation and putrefaction are peculiarly the effects of certain species of the low plants of which we write. Because the latter are so minute, they are not commonly known to be present, hence the popular idea that the processes are spontaneous, due to the nature of the material in which they occur. The French Count, Appert, early in the present century, working under the idea that the oxygen of the air is the active agent in these destructive changes, devised the method now so largely adopted of hermetically sealing fruits and meats. He fortunately hit upon the plan of expelling the air by heat, and of closing the vessel while the contents are hot. The result is what he hoped for, the indefinite preservation of the material. This explanation, though the only one known to many to-day, is absolutely false, as has been clearly shown in several ways. It is by no means necessary to exclude the air; it is only essential that living things adhering to the surfaces of the fruits, etc., and often contained in the air, be prevented from developing and multiplying in the substances. The heat destroys such as are in the vessel either on its own interior surface or on the fruit, and the closing while hot prevents the entrance of fresh germs. A can of such preserved material opened in the summer on the house top in the country might not spoil for many days or weeks. In such experiment care should be taken to keep the article to windward of the person for obvious reasons. In a warm time during winter success would be much more certain. Two years ago I found that milk taken direct from a cow in a heated glass fruit jar, with the simple precaution of previously washing the udder and adjacent parts and closing the jar as soon as possible, often kept sweet and fresh several days, many times as long as that left open in a clean milk room.

I now proceed to rehearse the results of some more careful and conclusive experiments which, it must be admitted, fully substantiate what has been said about fermentation and putrefaction being due to living organisms instead of to the air or any quality of the experimental material.

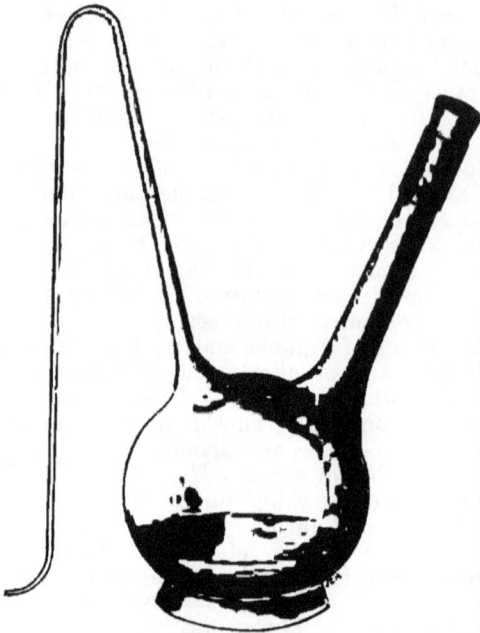

Figure 1 represents a little glass flask, invented by Pasteur, of France, for the purpose of making pure cultures of certain kinds of these minute organisms. The drawing was taken from one of the several in actual use by the writer, made at the chemical laboratory of the Industrial University. It will be seen to consist of a little bowl of the capacity of about one fluid ounce, and from this arise two tubes, one of which is surmounted by a short piece of rubber pipe, containing a glass stopper, the other long drawn out and bent downward, the descending portion being very fine, with an internal continuous cavity about the diameter of a medium sized horse-hair. The end is left open.

For our purpose we may take any fermentable or putrefactive infusion, as of hay or chicken broth. Filter to remove the solid particles, that we may better observe what does or does not take place, and fill the bowl of the flask through the stoppered tube.

We now hold the flask over a flame until the fluid boils, and the steam passes out of the larger tube, which we then close, having previously passed the stopper through the flame. The steam is permitted to rush out of the fine tube for a little time, after which we hold the fine open end in or just above the flame until the steam within the flask is condensed by cooling and the inrush of air ceases. We have now only to set the flask aside and watch results. If any change takes place, as a scum forming on the surface, a moldy growth on the glass, or a sediment settling on the bottom, it can readily be observed; if not, we may see that the clear liquid continues limpid and pure. Let us examine closely the conditions, supposing the temperature to be that of warm summer weather. We use a fluid which, freely exposed to the air or enclosed in a bottle without subsequent heating, would rapidly ferment or putrefy, becoming turbid and milky, with (if soup) a slimy scum forming on the surface, and withal an odor so characteristic that we shall know well enough what it means. By boiling in the flask, any living organisms in the liquid or attached to the interior of the vessel are killed. And by holding the minute opening in the flame—taking care that the glass is not melted—until after the condensation of the steam by cooling, there is afterward no strong inrush of air through the open tube. Now, all living things, no matter how large or how small, have solid parts, and are really heavier than the air, so that however minute, they cannot float in the latter, except as swept along by currents; hence, though by every alteration in temperature, without or within, there will be slow interchange of air, the minute solid particles of the external atmosphere are not likely to be carried up the fine tube and over into the liquid, which is therefore exposed to pure germless air. I have now such a flask, filled nearly two years ago with filtered chicken broth which is as clear and sweet as when first made, and this has been kept on a shelf in the laboratory, where the temperature has been very favorable to putrefactive changes, and where, from the use of the room, the air is unusually laden with organic dust. On the same shelf are other flasks of similar kind, some of which, after keeping for some months in the condition of the one described, have been unstoppered for a few minutes, and then closed again, with sure putrefaction as the result. Sometimes simply shaking is sufficient to start this process, showing well enough the preservation was not due to any want of susceptibility of the liquid to such change. If, however, we take the precaution to hold the larger tube over a flame, we may open and close it without such results, and in this way we may from time to time microscopically examine the liquid by taking a drop upon a previously heated glass rod. While the liquid remains limped and sweet, no organisms are found, but they always swarm in profusion when visible changes take place.

A successful experiment of this kind seems to prove two things: 1st, that the potential factor in fermentation resides neither in the substance itself nor in the pure air; 2d, that organisms do not spontaneously develope from germless material.

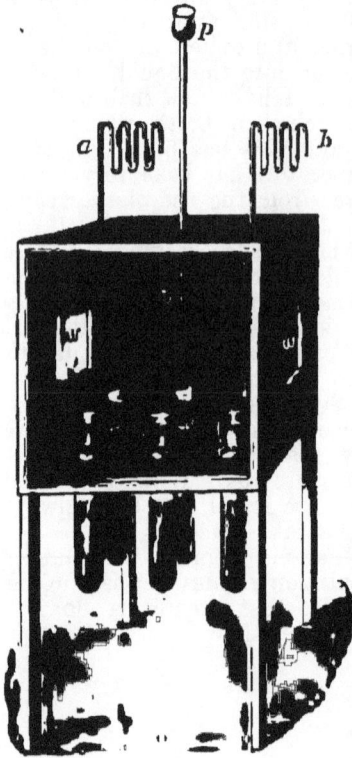

Fig 2.

Tyndale, of England, has taught us another mode of experimenting to gain answers to the same questions. He constructs a wooden box with a back door, glass front and side windows, and passes through the bottom several test tubes, air tight. Through a hole in the top fitted with a sheet of India rubber and a stuffing box of cotton wool wetted with glycerine, a long glass rod with a funnel above may be thrust to fill the test tubes, and two sinuously bent glass tubes pierce the top for the admission of air without dust, as explained in the account just given of Pasteur's flask. The interior surfaces are moistened with glycerine and the box closed. By passing a beam of sun, or brilliant artificial light, through the windows it can be readily ascertained when the floating matter has settled and the air is pure, after which the test tubes are filled with putrescible liquid, which is now boiled by heat applied externally. We notice that the boiled infusion is freely exposed to the air of the box as well as to the outer atmosphere through the bent tubes, and that the only precaution taken is to allow time for all minute solid particles to settle in the still air of the chamber to whose moistened surfaces they adhere. Hundreds of tests have been made in this way by Professor Tyndale with the most satisfactory and positive results, with very many kinds of putrescible substances, as natural animal liquids, infusions of flesh, the viscera of animals, of fish and of vegetables. The perfectly sweet and limpid filtered infusions remained in this condition for months, while outside of the moteless box in the same room they became putrid and offensive in some hours. It was sufficient at any time to open the back door of a box, if but for a moment, to secure these latter results in the experimental test tubes. In the clear, unchanged liquid living organisms were never found; in the ill-scented turbid ones, these always existed in countless numbers. What can be more instructive? What better demonstration could we wish of the non-occurrence under usual circumstances of spontaneous generation of living organisms, or of the stability without these of organic substances?

2. Diseases of Plants and Animals.

From the earliest times many of the diseases of man and of the domesticated animals have been known to be contagious, i. e. capable of being transmitted from a diseased to a healthy subject, while some that are now known to be propagated in this way were looked upon as peculiar visitations of Providence for supposed sin—the fact of the contagion not having been observed. There have also been from earliest history many speculations as well as careful studies, upon the nature of the poisonous or contaminating principle, and physicians, ever since their art has been practiced, have endeavored to find out remedies by experiment. Among•the diseases of the human skin that is happily much less prevalent now than formerly, the *itch* was the subject of extended and warm controversy from the twelfth to the eighteenth century among Arabian, Italian, French and German physicians and scientists. Some held that want of personal cleanliness was the sole cause and that the disease might spontaneously appear at any time and place, given only the proper conditions. Others found a minute, living, crawling, egg-laying inhabitant of the diseased skin, and attributed the ill effects directly to its work. Then came learned disputes as to whether this living thing did or did not originate from the filth, and whether it came as a consequence simply, or was verily the exciting cause of the disgusting malady. The literature upon the subject is very full, and if in those days they had possessed our aptitude for putting things in type, it would have required a library of books to contain the discussions.

Practicing the experimental method, some investigators at last put the living creature, separated from the *debris* of the diseased skin, on their own bodies, and watching its operations, established its true parasitic nature.

The agent, a mite, popularly called an insect, is propagated only from eggs laid by parent individuals, and thus maintains its specific identity. Its life-history having been satisfactorily made out, and its mode of operation fully made known, it is no longer a subject of dispute; but the history of the controversy is a valuable one in our present studies. Knowing the facts, we now smile at the absurdities of the opinions held, and the foolishness of the supposed proofs upon which these opinions were founded; yet these were not cruder nor farther from the truth than are many of the speculations and incredulities of our times concerning "diseased germs."

While an illustration of this kind cannot be taken as evidence in the case of bacteria, the same questions are now asked about, and the same objections made against, these latter organisms as the *cause* of disease. Yet careful investigators are to-day as able to handle bacteria, to see their shape, observe their manner of increase, ascertain their proper food material, and find out the substances inimical to their lives, as were Abenzoar and Bonomo in their time to similarly study the itch mite,—thanks to the improvement of the microscope, more than anything else, except the general acceptance of the experimental and inductive method of study.

The fact is, those competent to judge, now agree in holding that some severe contagious diseases of man and animals are directly due to bacteria, while many scientific investigators, whose abilities cannot be questioned, and who have abundant facilities and opportunities to learn and to know, claim that all diseases, readily communicable from the infected to the healthy, have their origin in the activities of these living organisms, and this opinion is certainly growing into wide, if not universal, acceptance. Other diseases, as fever and ague, not understood to be infectious or contagious, are also believed to be due to certain kinds of the same things.

Herein lies the greatest interest and highest importance attributed to these wonderful but excessively minute denizens of the, to common eyes, invisible world; and herein, more than anywhere else, rests the hope of attaining a scientific basis for medical practice, as well as a rational adoption of preventive measures against the ravages of disease. It will not do for the hobbyist to assume that all the ills that flesh is heir to can be traced to the corrupting work of "disease germs," for, as has been before pointed out, these minute things have no life functions which strictly separate them from other plants and animals. Their physiology is our physiology; they assimilate food material as we do, and, by virtue of this power, live as we live. The delicate and complicated machinery of the higher animal bodies may be put out of order in many ways, and by want of nature's harmony the normal, vital forces themselves may be the agents of disease.

But while there is no toleration for the hobbyist, those who have not investigated and cannot investigate for themselves should not hesitate to accept the testimony of capable specialists, when the latter find reason to assert that such and such a disease is due to the microscopic mischief-makers Their bacteria minuteness no longer prevents the demonstration of their presence, the tracing of their development, the detection of the actual effects and the experimental testing of results. There is now in certain cases just as good evidence that bacteria cause disease as there is that hawks destroy chickens, and the evidence is as inductively rigid in the one case as in the other. Even without microscopic examination, there is good reason to assert that the contagious principle, whatever it is, *grows*. Any chemical poison decreases in virulence by diffusion in a mass of inert matter, and soon loses its effectiveness; but it is the special characteristic of the poisons of which we write that a minute quantity soon infects the whole system of a large animal, and then the smallest drop of fluids from its body is sufficient to give origin to the disease in another animal and so on perpetually. Increase has taken place; there must have been growth; only living things grow; the microscope aids us to see what this living thing is; why should we doubt!

It is true that Dr. Lionel Beale proposed a theory of disease germs, which accounts for increase by assuming that degraded, yet not dead, parts of the animal body itself constitute the true contagion, and that every such living but degraded portion has the power in some way of over mastering the healthy—a speculation which, though brilliantly conceived and sustained at the time, has not been definitely supported by later investigations, nor held with assurance by any other authority, known to the writer; though, from what has just been said, the process is neither inconceivable nor *a priori* impossible, perhaps not even improbable. The only question is, "Do facts prove it?" There are indeed some experimental facts which seem to favor the idea, viz: the transmissibility of inflammation produced by chemical or physical means upon inoculations with the exuding serum. It may be said however that even in this there is the possibility of independent organisms being the real virus.

The first septic disease well worked out was that of *pebrine*, an infectious malady of silk worms in France. In 18;3 the production of coccoons was 57,000,000 pounds avoirdupois, and had doubled during the preceding twenty years with every prospect of continued increase; but during the next twelve years, owing to the ravages of the disease mentioned, the production fell to 9,000,000 pounds, a loss in 1865 alone of twenty million dollars. During those twelve years of national disaster, the disease had been seduously studied by all the scientific and professional skill of the times so far as the light of pathological knowledge then attained permitted, without practically beneficial results. In this state of things Pasteur was besought by the Minister of Agriculture to undertake the investigation, though this scientist had no special qualification for the work, except his natural talents and his acquirements through studies on the origin and effects of the micro-organisms in fermentations and putrefactions. Great numbers of minute corpuscles had been seen some years before in the bodily tissues of the diseased caterpillars, but no one attached any importance to them. Pasteur at once turned his attention to these microscopic bodies, proved them to be living organisms, studied their propagation and mode of development, became able to distinguish infected from healthy eggs, and in September 186; communicated his results to the French Academy of Science. His labors met with inattention or derision, so novel were they, and contrary to the learning of the time. To convince the skeptical and arouse the heedless, he selected fourteen lots of eggs in 1866 and wrote out what would be the fate of the moths hatched from them in 1867. These predictions, placed in the hands of a public officer in a sealed envelope, were exactly fulfilled in twelve of the fourteen cases, and produced the expected effect. His instructions were obeyed and France rapidly regained her great silk industry—one of the great triumphs of practical science in our triumphant century, and the foundation of an entirely new system of research in preventive and curative medicine.

Guided and profiting by Pasteur's former studies, Lister, then of Edinburgh, since of London, began in 1865 to perfect a system of anti-septic surgery, which, as introduced by himself and others, has been the means of inestimable progress in this difficult art, saving life and limb, as well as untold suffering to mankind. Unfortunately, either from the want of sufficient information or from placing too much stress on certain details, doubts are entertained by some fair-minded professional surgeons as to the correctness of the theory, though not upon the real progress that has been made. There seems good reason to believe such doubts will rapidly pass away, and that still further profit will be secured in the practical application of scientific knowledge. It is well known to veterinary surgeons that animals may be castrated by a sub-cutaneous rupture of the spermatic cords, followed by the death and absorption of the testicles, without mischief to the animal, provided the epidermis remains unbroken, but this provision is imperative. Gangrene is sure to follow if the exterior is punctured. Flesh bruises on our own bodies do not suppurate under similar conditions, and wounds made with a truly clean instrument heal by "first intention," if immediately closed and bandaged in such manner that living organisms are excluded. This result is more certainly assured by the proper application of an antiseptic dressing, as of glycerine, in which a small amount of carbolic acid has been stirred.

The etiology of no infectious disease is now so well understood as that known as "splenic fever," "anthrax" or "charbon" of the domestic animals and "malignant pustule" in man. It is said, between 1867 and 1870 over 56,000 horses, cattle and sheep, and 528 human beings perished from this alone in a single district of Russia, while the annual loss of live stock in France has been millions of francs. This is now practically subdued by prevention of infection and by vaccination, the history of which reads like a romance. The contributions thus made to medical practice opens up boundless visions of future welfare to man. Jennerian vaccination against (for ?) small-pox was the result of an acute but fortunate observation, one almost might say a lucky coincident; but Pastorian vaccination, as we may properly term it, is the outcome of methodical and precise scientific research, gaining knowledge step by step and reaching new facts by logical induction from the known. To this progress there have been many contributors, but among them it is not unjust to mention Koch, of Germany, and Pasteur, of France; the former for his splendid studies on the cause, the latter for the introduction of a practicable remedy, now adopted on a commercial scale in France and Austria.

The terrible disease is thoroughly known to be due to the inimical activities of a specific organism, *Bacillus anthracis*, described in another part of this paper. Within the last few years certain researches have been published which seemed to demonstrate that this was a physiologically varied form of a species of the same genus common in infusions of vegetable substances, as of hay; but very recent investigations make any such transformation questionable.

Pasteur's vaccination material is obtained by prolonged cultivation of the deadly Bacillus in a decoction of animal flesh kept at a certain temperature, in which the organism multiplies rapidly, but progressively loses its virulence. At the proper stage, the liquid swarming with this particular microphyte is used for the inoculation of animals which scarcely, or not at all, suffer in consequence, but are thereby protected from the virulence of the dreaded parasite in its usual condition, bringing almost certain death within twenty-four to thirty hours.

As an experimental illustration, Pasteur and his co-workers made a public test, at the request of a provincial agricultural society, and before a large assembly of interested persons and experts. Fifty sheep were taken from a flock, and from these twenty-five were ear-marked and vaccinated, May 3, 1881. This was repeated on the 17th of the same month. In the meantime the fifty sheep were kept together in the same flock, under the same conditions and influences. May 31 the whole fifty were inoculated with strong virus, and the sheep again turned loose together. Pasteur's predictions as to the result were fully verified the following day—twenty-three unvaccinated sheep being dead at two o'clock, and the two other during the afternoon, while the twenty-five vaccinated animals remained in perfect health, save that one purposely receiving an extra dose of the strong virus was sick for a few hours, then recovered.

Such satisfactory experiments having been often made, thousands upon thousands of domesticated animals have been vaccinated during the spring and summer of 1882, with material furnished by the renowned investigator, and again the financial interests of France owes to her illustrious son and to science a splendid recompense of reward.

Nor is the end reached here. The successful study of this disease and the production of a safe vaccination virus, will no doubt lead to similar results for other diseases, and not improbably furnish a scientific basis for that already practiced upon human beings as a preventive of small-pox. The fact is, a study of the disease of fowls commonly called cholera, helped in working out that of splenic fever, and especially led the way to the preparation of the vaccination material as now used. This disease of the domestic fowl occurring on both sides of the ocean is really of like nature, but want of space precludes further notice of it. Suffice it to say that it is readily transmissible, either through external puncture or wounds, or by taking the parasite with the food. This fact alone should soon lead to measures for its extermination.

In our own country, among other studies of this kind, much has been learned about yellow fever, undoubtedly due to a specific organism not yet satisfactorily determined; about diphtheria, in which the minute but dangerous enemy has been well recognized and carefully studied; about several forms of septicæmia or blood poisoning; and especially about a prevailing disease of swine ordinarily known as "hog cholera." Though "a prophet is not without honor save in his own country," let us hastily review the facts ascertained by Drs. Detmers and Law in regard to this last-named malady.

Hogs were entirely exempt in the United States from the disease until quite recent times, and though the original infection and the spreading from the first locality was not observed, many facts are known concerning its introduction in given areas and its subsequent mischief, where swine kept under the same conditions were robust and healthy before. Numerous speculations were made and theories advanced to account for the plague. By some it was attributed to the diet (too exclusively corn), by others to filth and unhygienic surroundings, and by others again to degeneration from improper breeding, etc.

The gentlemen named above, with others, were appointed by the Commissioner of Agriculture of the United States to investigate the disease, and, provided with the requisite means and apparatus, working separately, they satisfactorily proved the infectious character of the devastating malady. A fraction of a drop of the fluids from a diseased animal, placed beneath the skin of one in health by an inoculating needle, produced in a few days the characteristic affection, as did the diseased parts given in the food. They found by hundreds of post mortem examinations the lesions produced, and the course and progress of the injuries. They also found constantly present a specific organism, and thoroughly studied its morphological and physiological characteristics. They cultivated it outside the living body, in harmless fluids, and by further inoculations and examinations proved its direct agency in the production of the observed results. They showed how this organism was liable to be

transported by infested animals, sometimes other than hogs, by
water-currents, and, under certain limitations, through the air; they
found under what circumstances it was naturally destroyed, and
how preserved, and they succeeded in showing what preventive
measures could be used through general management, medicinal
treatment and vaccination. It was my good fortune to witness
many of the experiments of Dr. Detmers, and can personally vouch
for most of the above results, making, for myself, conviction irre-
sistible, that the immediate cause of the disease is none other than
the living organism always found in the blood and excretions of the
diseased animals.

Whether through the information thus derived, or from other
causes, the loss from this dreaded scourge has been conspicuously
less during the last few years. Certain it is, men are on the alert
to prevent, through intelligent endeavors, the infection of herds,
even though they do not, in words, admit the truthfulness of the
"germ theory."

The investigations of the author on "blight" of the pear tree and
other plants—the only facts yet made public of injuries to living
vegetation by bacteria,—have a scientific interest in consequence of
the readiness with which studies may be made. It is not possible
to cut pieces from the flesh of live animals and examine them under
the microscope in their living state; but this can be done with
plants. In the latter there are no sympathetic inflammatory processes, by which a healthy part suffers in consequence of injury to
some other part of the body; neither are there facilities for the distribution of microscopic parasites through the structure, as there is
in the blood currents of animals. We often say the sap "circu-
lates;" but no such thing really takes place. All the movements
of fluids in plants are processes of imbibition, or soaking through the
cell-walls, never in streams in tubular vessels. Hence disease infection is local, and the effects are comparatively very easy to make out.

In animals it is hard to determine whether the injury comes from
the destructive action of bacteria or from their mechanical obstructions, or the poisons they give rise to, etc.; in the pear tree such
queries are quickly answered, and one soon sees that death comes
from the fermentation of the cell contents. It is highly probable
that much can in this way be learned to help in the study of
similar maladies of man and the domestic animals.

The studies on blight have also a practical interest in the well
proved fact that the progress of the disease is slow, instead of the
rapidly spreading plague formerly thought. Careful examination,
once or twice a month, will usually suffice for the efficient excision
of all infected parts, and this, if properly done, is perfect protection
to the tree. The wounds must be quickly covered with paint or
other dressing.

The attested fact that the poisonous principle of certain plants,
as of poison ivy or oak (Rhus), is a living organism (T. J. Burrill,
Proceedings American Association for the Advancement of Science,
1882,) instead of a volatile chemical compound as heretofore supposed, is also a fruitful beginning in the study of the *origin* of infectious diseases. Of a similar nature is the ascertained fact that
a micro-organism found in the saliva of healthy human beings,

causes death when inoculated into a healthy rabbit, (Sternberg, National Board of Health, 1880).

Such facts offer no contradiction to the disease-producing agency 'of the observed living organisms; but only show that they may be harmless to one host and deadly to another, at least when first introduced. Many facts go to show that the physiological system may become inured to those poisons as well as to those of a chemical nature. As a man becomes habituated to the use of deadly doses of opium and arsenic, so in like manner he may resist after a time the pernicious effects of bacteria, though the latter continue to live and multiply in certain parts of his body. This probably explains the peculiar characteristics and effects of "Texas cattle fever." In this very characteristic disease of neat cattle the native animals of Texas do not suffer, while those recently introduced from abroad are almost certainly affected. Still more curiously, the healthiest Texan animals carry the contagion with them when transferred to the Northern States, where for a time the alarming death of cattle which have in anyway come in contact with those from the South excites the liveliest inquiry and most earnest investigation as to the cause. Until bacteria became known as true disease producers no satisfactory explanation could be made of the strange phenomenon.

What inscrutable mystery has been connected with that terrible scourge to human beings, leprosy! How it baffled the skill of the physician, and doomed the unfortunate to helpless and hopeless misery, making them loathsome though pitiable objects; life a wearisome burden to themselves, and their very being a menace to their friends! But the puzzle is solved. Through the microscope, the cause is visible as a self-perpetuating, parasitic plant, growing, propagating, destroying; yet subject to the same general surrounding conditions as other life-possessing things. Its terror is half gone from being known, and scientific medical treatment will banish the other half. Though by no means a disease of the ancient times alone, its uninterrupted career need no longer darken the lives of innocent sufferers superstitiously supposed to be under the curse of God's anger. The diseased need no longer be driven to the deserts or isolated in asylums for life in order to protect the community from the dreadful plague.

Without attempting to give a résume of what is known of the disease-producing effects of bacteria, but simply illustrations of this knowledge, mention is here made of only one more case, --the most dreaded and destructive of all the ills of the human body, consumption, (tuberculosis). Since 1865 this has been positively known to be transmissible from the diseased to the healthy, though not formerly so recognized. Up to the present year able investigators have sought in vain for the parasitic organism which all analogical facts strongly indicated must be the cause of the terrible disease. But skill and patient research have at last triumphed. By certain processes of staining, the first of which rewarded the labors of Dr. R. Koch of Berlin, the destroyer has been detected, and has now been seen by hundreds of expectant eyes. It inhabits diseased parts of the body, which may be in almost any organ as well as the lungs; it lives and grows and reproduces, slowly for its kind, but surely working its dreadful results. The disease is common to human beings and the domesticated animals, and is transmissible from one to another in either order. With such knowledge and the possibility now of definite and sure grounded study, the misery and suffering heretofore so calamitously common, must soon be beneficently controlled, and the world made brighter and happier to unnumbered thousands of its inhabitants. We cannot anticipate the results in regard to cure, but may comprehend means of prevention. There need be no more discussion about hereditary taint except as in direct conditions. Let it be well understood that the terrible disease is liable to be contracted from an animal or man suffering from it by any one predisposed to its attack, and let the modes of infection be made known, and instead of one-seventh of the human family perishing, as they now do after months and years of misery, the statistician, of the future may write one-seventy-seventh. Whatever alleviation there may be for those already diseased, the above is not too much to reasonably expect as to prevention.

3. BENEFITS.

The foregoing might lead one to the conclusion that bacteria were injurious and only injurious in their effects, but a little thought will soon convince us that this is far from the truth. They are primarily the agents of decay, yet in this very sense untold good comes from their activities. It is indeed no more startling than true to say that as at present constituted the organic world is indebted to them for its existence; without them man could not live upon the earth. The processes of nature run in a circle, the possibility and perfection of which depend upon the proper filling of every part. Green plants, the architects of the world, build solid structures out of gaseous and liquid materials, but must have these materials furnished. They create nothing. The supply being limited, this would soon be exhausted were there not some provision to prevent it. We have seen that organic materials, though dead, have no inherent tendency to decompose; kept free from the working of living things they endure forever, except when consumed by fire. How different the world would be without putrefaction and decomposition! Eliminate these processes now, and we should see the fallen trunks of trees accumulating in the forests as obstructions, instead of helping to form a rich vegetable mould for future growths and generations of forests; the mummied bodies of animals would remain for all time, like the grotesque cadavers of the old Egyptians, teaching history, but teaching only the sad one of the reign of death. It is well known that fertile soils have as an indispensable part of their composition a certain amount of partially decomposed vegetable matter; this is in truth the special characteristic of the surface soil as distinguished from the subsoil. Our field and garden plants are dependent on this rich upper stratum, and our valuable animals and ourselves are dependent on the plants; hence, following the circle round, we reach the knowledge of our indebtedness to the bacteria as agents of decomposition.

There is another way in which the fertility of soil has recently been shown to depend on bacteria. Green leaved plants require nitrogen as an essential part of their food, and they are only able to make use of this when in the shape of nitrates, or as combined with earthy or mineral substances. It has long been known that saltpeter (nitrate of potash) collects in caves. In times of need, it has been artificially obtained by washing earth in which there existed considerable amounts of organic matter undergoing decomposition, and to the surprise of the workmen such soil, after certain intervals of time, may be washed and re-washed without apparent exhaustion. So it came to be understood in time that in suitable caves the compound actually formed where it was found, not simply collected there through the percolation of water, etc. Upon heating a portion of such nitro-collecting earth, or treating it with chloroform, this peculiar power was observed to be lost, and by further investigation it was demonstrated that a microscopic living organism is the real agent in the work. Thinking no doubt in part of this, a well-known scientist has announced as the topic of a paper, "The soil a factory, not a mine."

We more directly make use of bacteria in many ways. The fermentations in which alcohol is the chief product, are commonly due to the yeast plants, closely allied to the bacteria; but the latter are rarely absent from brewers' yeast, used in bread-making. The so-called "salt-rising" is wholly dependent on the work of bacteria. Similar agents make for us vinegar and saur kraut; for the chemist, litmus; they "rot" flax; they "bleach" linen and cotton spread out on the dewy grass; they clean bones macerated in water by the anatomist; they purify the waters of rivers, cisterns and tanks, causing it in close reservoirs to "work," after which it is sweet and bright. It is said bacteria play an important part in the manufacture of cheese, and in the production of certain perfumes and flavoring extracts. They are found in germinating seeds and in the digestive organs of animals; but whether or not of real use in these cases, cannot be confidently stated. It is even probable that the wonder-working little creatures are used in medicine under cover of some other name and idea, the physician as well as the patient being ignorant of the real action of the dose. It is pretty certain that epidemics among noxious insects sometimes occur through the influence of bacteria. Considerable interest was recently taken in the apparent destruction, under certain conditions, of insects by yeast artificially applied to their bodies or given them to eat. But the results of experiments seemed to show that the yeast itself was not capable of injuriously affecting any insects, and it was left to conjecture whether or not there might be something else in some specimens of yeast cake or powder which, by its action, would explain the few known instances of the death of insects presumably from the effects the yeast tried. It has now been demonstrated by Prof. S. A. Forbes (*American Naturalist*, Oct., 1882,) that living chinch bugs are sometimes filled with enormous numbers of bacteria, and the observations so far made go to show that such bugs are decidedly unhealthy, and perish in great numbers before completing their full development. Certain it is that these pests of the grain fields do die off at times by some epidemic among them, as the writer can testify from his own observations. Hundreds of them, of all ages, have been found about harvest-time in little heaps, at the base of wheat plants, dead and dying, while the weather and other conditions were favorable for their life and development. The disastrous scourge to "silk-worms," before mentioned, shows that such diseases of insects are not only possible, but they really do exist. Pasteur suggests investigations to find out some such destroyer of the *Phylloxera* now so injurious to the grape vines in Southern Europe, and he is quite confident that success would crown proper efforts in this direction. There does seem to be in this a line of study of most excellent promise, not for one species of injurious insects only, but for the mastery by man over many of his minute but most dreaded enemies.

—

PART III. CLASSIFICATION OF BACTERIA,

AND A SYSTEMATIC DESCRIPTION OF THE SPECIES.

No one pretends to have made out a complete, natural classification of the *Schizophytæ* or bacteria; although several naturalists have embodied in systematic form, their ideas of the kinds and their relations, based on shape, development, motion, physiological effects, etc.; but all such classifications are acknowledged to be preliminary and more or less artificial. That, however, which proves to be most useful and apparently as nearly natural as any, is founded on the form of the cells and their organic association. It has been best worked out by Dr. Ferdinand Cohn, substantially as follows:

TRIBE I. SPHÆROBACTERIA.

Cells globular, or oval; size, very small, often less in diameter than .00004 in.; isolated, in pairs, or in chains of many articles, or when young and actively increasing in number, imbedded in gelatinous masses, called *zoogloea*, or when forming a pellicle on the surface of liquids, *mycoderma;* without true spontaneous motion, but oscillating in liquids by molecular trepidation.

But one genus, viz.: *Micrococcus.*

TRIBE II. MICROBACTERIA.

Cells elongated-oval or short-cylindrical: isolated, in pairs, or in chains of four, more rarely of many rather loosely attached, or sometimes in zoogloea; with active, spontaneous motions, when in nutritious liquids supplied with free oxygen.

One genus, viz.: *Bacterium.*

TRIBE III. DESMOBACTERIA.

Cells cylindrical, usually several times as long as wide, straight; isolated, or usually united in chains; often with spontaneous movements, but in the case of some species always without motion.

Genera: *Bacillus*, and with some doubt, *Leptothrix, Beggiatoa, Crenothrix.* The question about these three genera is whether they should be included among the *Schizophytæ* or referred to the filamentous *Algæ.*

TRIBE IV. SPIROBACTERIA.

Cells cylindrical, usually several times as long as wide; curved, or spirally wound; isolated, or united in chains of less or greater length; with active, spontaneous movements.

Genera: *Vibrio, Spirillum, Spirochæta.*

Besides the above, other species, multiplying by self-division, are referred to the following genera: *Sarcina, Ascococcus, Streptococcus, Myconostoc, Cladothrix* and *Streptothrix.* The old genus *Monas,* formerly including many species of low animals and plants, has been so modified that it now contains only a few forms of doubtful affinities.

It must be remembered that these genera are principally based on the shape and association of the cells, and that the latter are supposed to be in their adult condition. It may be that very arbitrary separations are made. It is quite possible, indeed, that the same specific organism assumes, under certain conditions, several of the proposed generic forms; but, it is at least probable that each genus named includes some species which do not change beyond the limits of the description. For instance, the spores of *Bacillus* would, from appearances alone, be classed as *Micrococcus* at first, and as *Bacterium* after a certain period of growth; but *Bacterium termo* never changes so much that the genus can be mistaken. Billroth goes so far as to claim that all the subjects of Cohn's classification, except, perhaps, those constituting the genera *Spirillum,* and *Spirochæta,* belong to a single species, which he names *Coccobacteria septica,* while Nägeli supposes there exists a small number of true species, each of which takes several forms.

With such differences of opinion among those most competent to judge, we cannot pronounce with any confidence upon the number or the actual characteristics of the species in existence; but that true species do exist we may feel well assured. When sufficient knowledge has been gained of their life histories, a natural classification can be arranged. In the meantime, such as we have already outlined must continue to be of much service, as it has been in the past.

Dr. Luerssen has arranged the following key to the genera:

I. Cells not in [cylindrical] filaments, separating immediately after division, or in couples [or chaplets] free or united in colonies (Zooglœa) by a gelatinous substance.
  A. Cells dividing in one direction only.
    a. Cells globular..*Micrococcus.*
    b. Cells elliptical or shortly cylindrical........................*Bacterium.*
  B. Cells dividing regularly in three directions, thus forming cubical families, having the form of little bags attached side by side, and consisting of 4, 8, 16, or more cells.................................*Sarcina.*
II. Cells united into cylindrical filaments.
  A. Filaments straight imperfectly segmented.
    a. Filaments very fine and short, forming rods....................*Bacillus.*
    b. Filaments very fine and very long................................*Leptothrix.*
    c. Filaments thick and long ....*Beggiatoa.*
  B. Filaments wavy or spiral.
    a. Filaments short and stiff.
      a. Filaments slightly wavy, often forming wooly flocks............*Vibrio.*
      b. Filaments spiral, stiff, moving only forward and backward......*Spirillum.*
    b. Filaments long, flexible, with rapid undulations, spiral through their whole length, and endowed with great mobility..........*Spirochæta.*

What follows is chiefly from Dr. Rabenhorst's *Kryptogamie Flora of Germany, Austria and Switzerland*, of which this part has been recently (1881) re-edited by Dr. G. Winter, of the University of Zürich, and which is translated from the German by the present writer. The original source of the most of the matter is Cohn's *Beitrage zur Biologie der Pflanzen*. As the work now stands the systematic description of species, herein given, is believed to be the fullest and most nearly complete of any in existence. In the English language the only similar publication is Dr. Sternberg's translation of Magnin's treatise.

Winter unhesitatingly classes the bacteria among the Fungi and includes in the latter all cellular Cryptogams (flowerless plants) devoid of chlorophyll. It is true that this rigid classification unnaturally separates some species certainly very closely allied; but, since the physiological effects are of prime importance in the practical study of these plants, this separation is less to be regretted. It must certainly serve a useful purpose to present together those which, through the want of chlorophyll, are dependent on the assimilated products of other plants and animals for nutriment, and which thus agree in being agents of destruction in organic matter.

I have added to those in Winter's work such further species as seem to be well established, as well as some doubtful but often quoted names—the latter mostly by Hallier, and are copied from Magnin. The following are herein described as new species:

Micrococcus amylocorus, the "blight" of pear trees, etc.

Micrococcus toxicatus, the "poison" in species of *Rhus* (Poison Ivy, etc.)

Micrococcus insectorum, in diseased chinch bugs and supposed to be the cause of an epidemic destruction of these insects.

I have also felt obliged to name anew the organism found by Dr. Detmers and others in diseased pigs, for though there is a general agreement that the species belongs to *Micrococcus*, no one has published a name for it thus classified. It may now be known as *Micrococcus suis*. For the organisms causing the disease of the common fowl usually known as "chicken cholera" I have proposed the name of *Micrococcus gallicidus*. Several species have been recorded without specific names, because properly published names for them are not known by me. Thus, while some write *Bacillus tuberculosis* for the recently discovered species in this disease, I am not informed that the name as such has been published by Koch or any one else according to established usage. Still this may be the case and a new name should not be given. I have therefore simply written *Bacillus of tuberculosis*.

All notes and additions by myself to the German text are inclosed in brackets. In the translation I have endeavored to express the evidently intended meaning of the author rather than to make a literal rendering of the wording.

PLATE I.

FROM "MICROSCOPICAL JOURNAL."

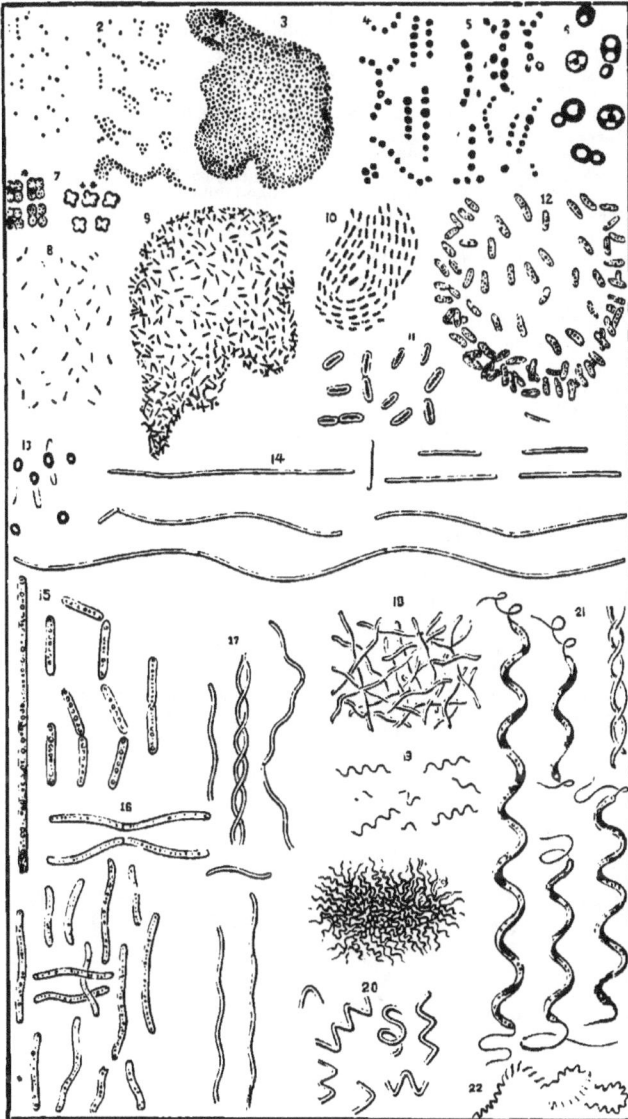

DESCRIPTION OF PLATE I.

[Fig. 1.—Micrococcus prodigiosus (Monas prodigiosa, Ehr.) Spherical bacteria of the red pigment, aggregated in pairs and in fours, the other pigment bacteria are not distinguishable with the microscope from this one.

Fig. 2.—Micrococcus vaccinæ. Spherical bacteria, from pock-lymph in a state of growth aggregated in short four to eight-jointed straight or bent chains, and forming also irregular cell-masses.

Fig. 3.—Zoöglœa-form of Micrococcus, pellicles or mucous strata characterized by granule-like closely set spherules.

Fig. 4.—Rosary chain (Torula form) of Micrococcus ureæ, from the urine.

Fig. 5.—Rosary chain and yeast-like cell-masses from the white deposit of a solution of sugar of milk which had become sour.

Fig. 6.—Sacharomyces glutinis (Cryptococcus glutinis, Fersen), a pullulating yeast which forms beautiful rose-colored patches on cooked potatoes.

Fig. 7.—Sarcina spec, * from the blood of a healthy man, * * from the surface of a hen's egg grown over with Micrococcus luteus, forming yellow patches.

Fig. 8.—Bacterium termo, free motile form.

Fig. 9.—Zoöglœa-form of Bacterium termo.

Fig. 10.—Bacterium, pellicle, formed by rod-shaped bacteria arranged one against the other in a linear fashion, from the surface of sour beer.

Fig. 11—Bacterium lineola, free motile form.

Fig. 12.—Zoöglœu-form of B. lineola.

Fig. 13.—Motile filamentous Bacteria, with a spherical, or elliptical highly refringent "head," perhaps developed from gonidia.

Fig. 14.—Bacillus subtilis, short cylinders, and longer, very flexible motile filaments, some of which are in process of division.

Fig. 15.—Bacillus ulna, single segments and longer threads, some breaking up into segments.

Fig. 16.—Vibrio rugula, single or in process of division.

Fig. 17.—Vibrio serpens, longer or shorter threads, some dividing into bits, at * two threads entwined.

Fig. 18.—"Swarm" of V. serpens, the threads felted.

Fig. 19.—Spirillum tenue, single and felted into "swarms."

Fig. 20.—Spirillum undula.

Fig. 21.—Spirillum volutans, * two spirals twisted around one another.

Fig. 22.—Spirochæta plicatilis.

All the figures were drawn by Dr. Ferdinand Cohn with the immersion lens No. IX, of Hartnack Ocular III, representing a magnifying power of 650 diameters.]

KEY TO THE GENERA OF SCHIZOMYCETES.*

SYNOPSIS OF THE GENERA.

Micrococcus.—Cells globular or oval-elliptical, motionless,+ dividing only in one direction, isolated or united in chains or in zooglœa.

Ascococcus.—Cells globular, united in irregular families, which are often lobed and surrounded by a capsule of firm, cartilage-like jelly.

Cohnia.—Cells globular, imbedded in a single peripheral layer of jelly, which is spherical and hollow, or at a later stage irregularly bladder-form; the forms like nets are broken through.

Sarcina.—Cells globular, dividing in two or three directions; daughter cells small, united in solid or tabular families, mostly in fours, or some multiple of four.

*[Dr. Winter adheres to the opinion that the Fungi and Algæ are distinct classes of plants, and includes the bacteria among the former. For this reason the term Schizomy-cetes is chosen rather than Schizophytæ.]

[+ i. e., having only "Brownian" movement—not swimming freely from place to place.]

—9

Bacterium.—Cells short cylindrical, or long elliptical, or fusiform, with rapid movements; otherwise as in *Micrococcus.*

Bacillus.—Cells elongated cylindrical, mostly united in filaments dividing transversely; forming spores.

Leptothrix.—This doubtful genus is characterized by the very long, slender, unbranched and apparently unarticulated filaments.

Beggiatoa.—Filaments very long, rather thick, mostly indistinctly articulated, actively vibrating, containing highly refractive granules.

Cladothrix.—Filaments very slender, indistinctly articulated, pseudo-branched.

Myconostoc.—Filaments very slender, bent and twisted through each other, imbedded in globular, jelly-like masses.

Spirochæta.—Filaments long and very slender, with numerous close spirals; movements lively.

Spiromonas.—Cells flattened, spirally twisted.

Spirillum.—Cells cylindrical, with a single curve or spirally wound, mostly with a cilium at each end.

APPENDIX.

To the *Schizomycetes* are appended the following allied genera whose systematic position is still doubtful:

Sphærotilus.—Cells arranged end to end in a colorless, gelatinous sheath, forming long threads and flocks.

Crenothrix.—Cells united in filaments surrounded with a sheath.

SCHIZOMYCETES.

The *Schizomycetes* or cleft-fungi are one-celled plants, which multiply by repeated division in one, two or all three directions of planes, and which also abundantly reproduce themselves by spores formed within the cells.

They live isolated or united in various ways, in fluids and in living and dead organisms, in which they induce decompositions and various—but not alcoholic—fermentations.

MICROCOCCUS, Cohn.

(Beiträge z. Biol. d. Pflanzen, I Bd. 2 Heft, p. 151.)

Cells colorless or slightly tinted, globular or oval-elliptical, motionless, dividing only in one direction. The daughter cells either soon separate from each other or remain in pairs or are united in chains or form zooglœa. Not certainly known to produce spores. [The molecular oscillations of these minute bodies must not be mistaken for spontaneous movements.]

The species of *Micrococcus* are not readily discriminated. The supposed distinct kinds show in form and size little or no differences and there remains only their chemical activities as the means of distinguishing them, which may be managed with considerable completeness.

A.—*Pigment forming Micrococci.*

M. Prodigiosus, Cohn.

Synonyms. *Monas prodigiosus.* Ehrb. (Monatsbr. der. K. Akad. d. Wissensch z Berlin 1848.) *Palmella prodigiosa.* Mont. Bulletin de la Soc. Nat. et cent d' Agricult d_Paris, 2 Sér. VII. p. 727; *Zoogalactina imetropha.* Sette (Mém. venezia 1824. *Bacteridium prodigiosum.* Schroeter Beitr. z Biol. I. 2 Heft. p. 109.)

Exsiccata: Rabenhorst, Algæ, 200; Thümen, Mycothcca Universalis, 1590.

Cells globular or oval, colorless, .00002 to .00004 in. in diameter; at first rose-red, then blood-red, finally growing pale; forming a slimy substancel

Micrococcus prodigiosus is the organism which, as known for some time, produces the singular phenomenon called in earlier times the blood of bread, the blood of the Host, etc. It forms at first minute rose-red points, and little heaps, which, becoming larger, form circumscribed, roundish, deep-red spots; afterward these spots run together and spread out, at the same time become dripping with a blood-red semi-fluid material. This consists of a red, gelatinous mass in which millions of cells of the Micrococcus are imbedded. The latter are colorless; they give out the pigment to the jelly. The coloring matter, in its chemical and physical characteristics, very much resembles fuchsin. It is insoluble in water, but completely soluble in alcohol, and this solution evaporated to dryness and again dissolved is orange-red. On the addition of acids it becomes bright-red; with alkalies yellow. In the spectroscope it shows with others a characteristic broad absorption band in the green. *Palmella mirifica*, Rabh., can scarcely be different.

M. luteus, Cohn.

Synonym: *Bacteridium luteum*, Schroeter (l. c. p. 119 und 126).
Exsiccata: Thümen, Mycothcca Universalis, 1400.

Cells elliptical, a little larger than those of *M. prodigiosus*, with highly refractive contents; forming light-yellow drops on a solid substratum, at first the size of a poppy seed, later the half of a grain of pepper, and finally drying up into flat, shield-form, little bodies. This species forms, on the nourishing fluid, a thick yellow skin which becomes wrinkled if the development is luxuriant.

On boiled potatoes, etc.

The coloring matter is insoluble in water; it is not changed by sulphuric acid or alkalies

M. aurianticus, Cohn.

Synonym: *Bacteridium aurantiacum*, Schroeter (l. c. p. 119 und 126.)
Exsiccata: Thümen. Mycothcca Universalis, 1700.

Cells oval, .00006 in. long; forming, on a solid substratum, orange-yellow drops and spots, which finally spread into a uniform coating; on nutritive fluid it forms a golden-yellow layer.

On boiled potatoes and eggs.

Coloring matte soluble in water.

M. ulvus, Cohn.

Exsiccata: Rabenhorst, Algæ Europeæ, 2501.

Cells globular, .00006 in. in diameter; at first forming rust-red conical little drops .02 in. in diameter, these enlarge and finally appear as a broad mass of slime.

On horse-dung.

M. chlorinus, Cohn.

Cells globular (?); forming golden or verdigris-green slime masses, or a verdigris-green layer on fluids which gradually becomes colored throughout.

On boiled eggs.

The coloring matter is soluble in water; it does not turn red with acids.

M. cyaneus, Cohn.

Synonym: *Bacteridium cyaneum,* Schroeter, (l. c. 122, und, 126.

Cells elliptical, producing on slices of potato an intense blue coloration which also penetrates the substance and even shows on the opposite side of the slice. On nutritive fluids zoogloea are formed which are at first colorless, then bluish-green and finally an intense blue.
On boiled potatoes.

The coloring matter is soluble in water; the solution is at first verdigris-green, afterwards as a rule pure blue. It is made by acids an intense carmine-red and changed again by alkalies to blue or gall-green. In the spectroscope it shows no absorption-line but only a darkening of the less refractive parts.

M. violaceus, Cohn.

Synonym: *Bacteridium violaceum,* Schroeter (l. c. p. 157.

Cells elliptical, larger than those of *M. prodigiosus,* in little bright violet-blue drops of sl me which enlarge (to one-fourth of an inch in diameter) and run together into spots.
On boiled potatoes.

B.—Micrococci producing fermentations.

M. ureæ, Cohn.

Cells globular or oval, .00005 to .00008 in. in diameter, isolated or united in chains, or forming zoogloea on the surface of liquids.
In urine.

Micrococcus ureæ is the ammoniacal ferment. When fresh urine is allowed to stand at a proper temperature (30° C.), it loses in a few days its acid reaction, becomes neutral and finally alkaline, showing signs of fermentation. The urea is changed into ammonium carbonate, while at the same time ammonio-magnesium phosphate is precipitated. This fermentation follows only when the Micrococcus is developed in the liquid.

M. crepusculum, Cohn.

Synonym: *Monas crepusculum,* Ehrb. (Infus, p. 6, t. I, Fig. 1).
Exsiccata: Rabenhorst, Algen Europa's, 2502.

Cells globular or short-oval, very small, scarcely .00008 in. in diameter, isolated or forming zoogloea.
In and on various infusions and foul liquids.

This common Micrococcus appears with *Bacterium termo* in all foul substances and infusions.

M. candidus, Cohn.

Forms on slices of boiled potato snow-white points and spots.

M. of nitrates.

Schloesing and Muntz have communicated to the French Academy of Science the results of studies on the formation of nitrates in ordinary soil, and they prove that a minute, globular, or slightly elongated organism, is the cause of the phenomenon so common in nature, and of such vital importance in the fertility of the soil. *Comp. Rendus,* T. 89, pp. 891 and 1190. More recently, Guvon and Dupetit have shown that these same nitrates are decomposed by another *Micrococcus,* possibly *M. ureæ.* (l. c. T. 95, p. 644.

C. - Disease producing Micrococci.

M. Vaccinæ, Cohn.

Synonym: *Microsphæra vaccinæ,* Cohn (Virchow's Archiv. LV.)

Cells globular .00002 to .00003 in. in diameter, isolated or in pairs or united in chains or masses, or forming zoogloea.

In fresh lymph of vaccine vesicles of the cow, and of man, as well as in the pock-postules of variola (small pox.)

Micrococcus vaccinæ must be accepted, after the many reliable investigations, as the effective element in vaccine virus. It is the carrier of the contagion of small pox. By filtering the lymph the solid constituent can be separated from the fluid. When the latter is used for inoculation no effect is produced, while the former induces the formation of the pock-vesicles. But that the micrococci and not the lymph cells are the active constituent of the solid residue appears from the fact that when vaccine virus is exposed to the air for a time it becomes less and less effective. Such virus finally putrefies, and with the increase of putrefaction the micrococci correspondingly disappear, displaced by rot-producing *Bacteria*.

M. diphtheriticus, Cohn.

Cells oval .000013 to .000004 in. long, single or united in chains or forming variously-shaped masses and colonies.

In the so-called diphtheritic membrane found especially on the mucus surfaces of the throat, pharynx, windpipe, etc., but also of those of the sexual and digestive organs, as well as in wounds,' etc.

This organism is of very great pathological importance: for the infection spreads from its place of origin through the lymph vessels and their enclosing tissue, later into the connective tissue, the kidneys, the muscles, and finally the organisms gain entrance to the blood vessels, where they cause the greatest disturbances. They plug the capillaries and cause them to burst. The thinner bones and cartilage are destroyed through this same process. The contagious properties of the fungus is also very great.

M. septicus, Cohn.

Synonym: *Microsporon septicum*, Klebs. (zur. patholog. Anat. der Schusswunden, 1872.)

Cells globular .00002 in. in diameter, united in chains or masses or forming zooglœa.

In wounds; generally with all the kinds of disease known as *Pyæmia* and *Septicæmia*.

In the various states of suppurating and putrefying in the living body, in blood-fermentation and blood-poison, this micrococcus plays an active part. Whether all the manifold phenomena are called forth by micrococcus septicus or more kinds take part, is questionable. In wounds we find the micrococci in the fresh pus, in which they multiply rapidly and bring on inflammation and fever, destroying the tissues and penetrating deeper and deeper. They gain entrance to the blood vessels and cause obstructions and festering; similar phenomena occur in the lungs and liver.

[M. bombycis, Cohn.

Synonym: *Microzyma bombycis*, Bechamp (Comptes Rendus, tome 64, 1867, p. 1045.) Exsiccata, Umlant w.; Thümen Mycotheen Universalis 1799.

Cells oval .00002 in. in diameter, single or in chains. .

In the gastric juice and intestines of silk-worms, producing in them the so-called "Schlaffsucht" [pébrine], an infectious disease from which the animal after a short time dies.

There are probably many other infectious diseases, as cholera, measles, scarlatina, typhus fever, etc., which are due to *Bacteria*. Reliable observations are wanting upon them.

[M. of traumatic erysipelas.

This disease has been abundantly proved to be contagious and in many cases very virulently so. There have also been very numerous observations upon the existence of globular organisms in the excretions, but it has not been known until recently whether these living things were active agents in, or merely accompaniments of, the disease. The investigations of Orth, Recklinghausen and Lukomsky, corroborated by Koch and others, seem to demonstrate the fact that a *Micrococcus* is the real cause of the occurring pathological changes and the active element in the contagion. Puerperal fever is probably due to the same organism.

This *Micrococcus* is described as globular, isolated or in chains, without motion.]

42

M. of croupous pneumonia.

Many investigators agree in finding minute organisms in the excretions of inflamed tissues affected with this disease. Friedländer of Berlin has recently (1882) made careful researches, which seem to establish the fact that the living organisms exist and multiply within the tissues and blood vessels and not simply in the exudations. The description is as follows: Cells ellipsoidal, .00021 in. wide, .00801 in. long, usually in pairs, sometimes in chains or spread out in a film, aggregated in colonies in the lymphatic vessels. Lancet (London) March 4, 1882.]

[M. amylovorus, Burrill.

Cells oval, single or united in pairs, rarely in fours, never in elongated chains, imbedded in an abundant mucilage which is very soluble in water; movements oscillatory; length of a separate cell .00004 to .000056, in.; width, .000028 in.; length of a pair .00008 in.; of four united about .000012 in.

The cause of "blight" in plants, especially of the pear tree (fire blight) and of the apple tree (twig blight and sun scald.) The organism gains entrance to the living tissues through wounds or punctures and produces butyric fermentation of the starch stored in the cells. The disease is transmissible by artificial inoculation. (Tenth Report Illinois Industrial University. Transactions American Association for the Advancement of Science, 1880.)]

[M. suis, (Detmers) Burrill.

Cells globular, or elongated and more or less contracted in the middle, single or in pairs or chains of many articles; .000028 to .000032 in. in diameter (Detmers).

Found in the blood and other fluids of pigs suffering with swine plague or "hog cholera." Dr. Detmers at first classed this organism with the Bacilli and named it Bacillus suis. But Detmers, Mégnin and Salmon essentially agree in giving the characteristics of Micrococcus, and with this some observations of the author accord. Detmers describes a peculiar contraction of the middle portion of elongated forms rendering the sides concave, and in this condition he finds them in zooglœa and in chains. Others, according to the same authority, as well as Mégnin and Salmon, are spherical. It seems well established that the Micrococcus is the real contagious element in the disease. (Annual Reports United States Department Agriculture 1879-80.)]

[M. rox catus, Burrill.

Cells globular, single and in pairs, rarely in chains of several articles; .00002 in. in diameter; movement oscillatory only.

Apparently parasitic on species of Rhus, and constituting the poisonous principle in these plants. Reaching the human skin, the organisms penetrate in some way its tissues, and multiplying there induce the peculiar inflammation which takes place. (Some Vegetable Poisons. Trans. American Association for Advancing Science, 1882. Am. Mic. Journal, October 1882, p. 192.]

M. insectorum, Burrill.

Cell obtusely oval, isolated or in pairs, rarely in chains of several articles; .000022 in. wide and .000027 to .00004 in. long, usually about .000032 long; movements oscillatory only; forming zooglœa (?). In the digestive organs of chinch bugs (Blissus leucopterus).

Professor S. A. Forbes discovered this minute organism, in 1882, infesting the intestines of chinch-bugs, which evidently suffered thereby 'S. A. Forbes, American Naturalist, Oct. 1882.' Very many of the insects, especially the older ones, were found to harbor the parasite in great numbers, and at certain times very many of the bugs were found dead and dying before reaching maturity. It is well known that these pests of the grain fields do perish at certain periods by some epidemic disease, from which few individuals escape, and there is every reason to believe that the organism found by Professor Forbes causes the destruction. He finds it can be successfully cultivated in beef-broth, and the possibility is thus apparently within reach of artificially introducing and spreading the disease. If so, a most important step has been made in economical entomology and scientific agriculture.

The organism is somewhat similar to, but not identical with, Micrococcus bombycis, the "disease germ" of the silk worm, which was so fatally destructive to the silk industries of France, and which became the subject of the successful studies of Pasteur.

On the stems and leaf-sheaths of maize injured by the bugs a *Micrococcus*, supposed to be the same, was found imbedded in firm zoogloea masses.

The form of the organism approaches the typical shape of *Bacterium*, being between oval and short cylindrical, with rounded ends; otherwise the characteristics are those of a true *Micrococcus*.]

(M. of fowl cholera.

While all who have carefully studied this disease agree that the contagious element consists of minute globular granules, capable of self-multiplication, it appears that no one has either named the organism or given a scientific description of it. Toussaint (Compt. Rend. xci (1880) p. 301) supposed he had sufficient proof of the identity of this disease of the domestic fowl and septicæmia, or blood-poisoning, and that in both cases the living organism believed to be the active agent, is the same. But Pasteur (l. c. p. 457) pronounces them similar in appearance but quite distinct in effect. Salmon (Report Dpt. Agriculture (U. S.) 1880, p. 401) confirms the infectious nature of the disease, and the virulence of the multiplying granules, but is inclined to attribute their origin to the transformations of the animal protoplasm (p. 439.) There can scarcely be a doubt but that the so-called granules constitute a true species of the present genus. It may be called *Micrococcus gallicidus*.]

DOUBTFUL SPECIES.

M. griseus, Winter.

Synonym: *Bacterium griseum*, Warming.

Cells nearly globular or oval, colorless; .00010, to 00016 in. long, with those dividing .00024 to .00028 in. long, .000072 to .00010 inches wide.

In infusions of fresh and salt water.

Since according to Warming this form appears only in the motionless state, (and then forms no zoogloea) and since the shape of its cells corresponds better to the genus *Micrococcus* than to *Bacterium*, I have placed the fungus in the former genus.

M. ovatus, Winter.

Synonyms: *Panhistophyton ovatum*, Lebert. (Ueber die gegenwartig herrschende Krankheit des Insect der Seide in; Jahresbericht ueber die Wirksamkeit des Vereins zur Beförderung des Seidenbaues für die Provinz Brandenburg im Jahre 1856-57, p. 28 und folgde.)

Nosema bombycis, Naegeli. (Botan. Zeitg. 1857, p. 760; Flora 1857, p. 684.

Cells oval, twice as long as wide, rounded at the ends; .00016 to .0002 in. seldom 00024 in. long, .00003 to .00012 in. (mostly .00010 in.) thick, isolated in pairs or united in little masses.

In various organs of the silk-worms, their pupæ, and winged forms.

It is questionable whether the described cells belong to the *Schizomycetes*. They were first discovered by Cornalia in Mailand and designated corpuscles; according to him they are also found, though very scarce and more incidentally, in the blood of healthy caterpillars. Later these small bodies have been recognized as the cause of the epidemic disease of the silk-worm known as "gattine."

As the cells in shape and want of motion agree very well with *Micrococcus*, I have introduced them here.

[The eight following, by Hallier, are given without much confidence in the correctness of his observations and interpretations. These are taken from Magnin's "*Bacteria*."]

M. of animal variola, Hallier.

Small, endowed with lively movement, furnished with a very delicate appendage, sometimes united in the form of little elongated rods; found in spontaneous or inoculated pustules, in the lymphatic canals and the ganglia of animals attacked with variola.

M. of rugeola, Hallier.

Very small, without color, often with a caudal prolongment; in the sputa and blood of the sick.

M. of scarlat'na, Hallier.

Free or in colonies, either on the surface or in the interior of blood corpuscles, or in chains.

M. of ep demic diarrhœa, Hallier.

In intestinal matters with vibrios, cells and monads.

M. of exanthematous typhus, Hallier.

Relatively large brown, endowed with rapid movement, sometimes in chains, in the blood.

M. of intestinal typhus, Hallier.

Very small, without movement in the blood; larger forms with quick movements, furnished with contractile appendages; in dejections. Similar Micrococci are found in cholera diarrhœa, but in less number.

M. of glanders, Zürn.

Cells free or adhering to the blood corpuscles, or even penetrating in their interior; sometimes in chains; in the blood. Very numerous, endowed with rapid movement; in the lymphatic ganglia, in the mucus of the frontal sinus and in chancroid ulcers.

M. of syphilis, Hallier.

Numerous, colorless, free or in globules; in gonorrhœa, the primitive ulcer, and in the blood of subjects with constitutional syphilis.

ASCOCOCCUS, Cohn.

(Beitr. z. Biol. I, Bd. 3, Heft. p. 151.)

Cells colorless, very small, globular, united in larger or smaller globular or irregular families in shapeless masses. Families often lobed, the lobes again incised, inclosed in a firm cartilage-like gelatinous capsule of rounded form.

The value of Cohn's genus *Ascococcus* is to me as questionable as that of Billroth's genus of that name; it is also doubtful whether the two are identical. Possibly *Ascococcus* is only a stage of development of *Micrococcus*.

A. Billrothii, Cohn.

Families lump-like, .00080 to .00.'40 in. in diameter, inclosed in a capsule from .00004 to .00900 in. thick; covering the surface of the liquid in a thick, flaky layer.

On a solution of acid ammonium tartrate, forming a pellicle.

The colonies consist of a sharply defined cartilage-like gelatinous envelope, in which either one or more families are inclosed. The families are of very different size and shape, solid, composed of numerous, exceedingly minute, globular cells. This fungus induces in its nourishing fluid singular fermentations. It produces of ammonium tartrate, butyric acid and butyric ether, and changes the originally acid solution to alkaline while free ammonia is given off.

45

COHNIA, Winter.

Synonym: *Clathrocystis* (Henfry), Cohn (Rabenhorst's Algen Europas, No. 2318.

Cells globular; inclosed in a single, peripheral layer of jelly, forming a spherical, or later an irregularly shaped bladder or sack, the walls of which finally break up in a net-form manner. Increase takes place by repeated separation of the cells in twos; of the families through the folding in and lobing of the daughter families.

The genus *Clathrocystis*, Cohn, is removed according to my idea of the distinction between the *Algæ* and *Fungi*. Since this generic name was primarily adopted for an alga (*Cl. aeruginosa*, Henfry) it is advisable to retain the name for it and to make this species belonging to the *Fungi* the representative of a new genus, to which I have given the name *Cohnia*, in honor of Dr. F. Cohn, of Breslaw, highly merited for his investigations upon the *Schizomycetes*.

C. roseo-persicina, Winter.

Synonyms: *Protococcus roseo-persicina*, Kütz. (spec. Alg. p. 196); *Pleurococcus roseo-persicina*, Rabh. (Flora Europ. Alg. III, p. 28; *Microhaloa rosea*, Kütz. (Linnæa VIII, p. 311); *Bacterium rubescens*, Lankaster. (Quart. Jour. Mic. Le. XIII, New Series, p. 408, Pl. XXII, XXIII); *Clathrocystis roseo-persicina*, Cohn, (Rabh. Alg. Europ. No. 2318 and Beitr. z. Biol, I. Bd. 3, p. 157, Taf. VI, Fig. 1-10.)

Exsiccata: Rabh. Algen Sachsens, etc., 986 und 2318; Wartmann und Schenk, Schweiz' Kryptog. 343.

Cells globular, oval, or through mutual pressure, polygonal; from rose to purple-red; .00010 in. in diameter. They form at the beginning little solid families in which the single cells are joined by gelatine, while the whole family is inclosed in a gelatinous envelope. Later, there forms a larger, globular or oval, finally irregular, hollow body, attaining a diameter .02640 in., and filled with a watery liquid. In this the cells are arranged in a single peripheral layer. These bladders are often torn or perforated, becoming elegant nets, which finally break up into irregular patches and shreds.

In swamps, swimming on the surface or among algæ and duckweeds; often also in aquaria, in which algæ, etc., are decomposing.

The only species of this genus, so far known, is remarkable for its red coloring matter' which is essentially different from that of *Micrococcus prodigiosus*, and is known as "bacterio-purpurin." This is insoluble in water, alcohol, etc., is changed by hot alcohol into a brown substance, and is otherwise characteristic by its optical deportment. It shows through the spectroscope the same strong absorption band in the yellow; weaker ones in the green an I blue, as well as the darkening of the more refrangible half of the spectrum. The single cell is surrounded by a firm, almost cartilaginous membrane; their contents are, when young, homogeneous, but when older there appears within, one or more opaque granules, which are none other than metallic sulphur.

SARCINA, Goodsir, (Extended).

Cells globular, dividing in two or three planes; daughter cells a long time united, forming little solid or tabular families, which are often again united into larger colonies. As a rule, the families consist of four, or some multiple of four cells.

S. ventriculi, Goodsir.

Synonyms: *Merismopedia Goodsirii*, Husem (le anim. et veg., p. 13). *M. ventriculi*, Robin (Hist. des Veget. paras., p. 331, T. I, Fig. 8, et T. XII, Fig. 1).

Exsiccata: Rabh. Algen 600. Wartmann und Schenk, Schweiz. Kryptog. 247.

Cells globular, four, eight, sixteen or a few more united into little cubes with rounded corners—the parts in contact flattened; cells of the colony attaining a diameter of .00016 in.; colonies strung together by the partition walls of the cells; again united into larger masses. Cell contents greenish, yellowish or reddish-brown, somewhat polished.

In the stomach of healthy and diseased man and higher animals; also sometime occurring in other parts of the body.

S. urinæ, Welcker.

Synonym: *Merismopedia urina*, Rabh. (Flora Europ. Alg. II, p. 59).

Cells very small, .00001 to .00008 in. in diameter, eight to sixty-four united in a family; eight united cells .00008 to .00012 in., sixty-four cells .00016 to .00020 in. in diameter.

In the bladder.

S. littoralis, Winter.

Synonyms: *Erythroconis litoralis*, Oersted; *Merismopedia litoralis*, Rabh. (Flora Europ. Alg. II, p. 57).

Cells globular or, when ready to divide, oval; .00012 in., seldom .00008 in. or more in diameter; united in families of four, six, eight, etc., which are again grouped in larger colonies (as many as sixty-four 4-parted cells). Cell contents colorless, but there is in each cell one to four red granules of sulphur.

In putrid sea-water.

S. Reitenbachii, Winter.

Synonym: *Merismopedium Reitenbachii*, Caspary (Schriften der physikal. ökon. Gefell-schaft zu Königsberg, XV, 1871, p. 101, T. II, Fig. 7 to 15).

Cells globular, or when ready to divide, oval or elliptical, .00006 to .00010 in diameter, dividing ones attaining .00016 in. long; rarely single or in twos or threes, usually four or eight, or more often sixteen or more united. Cell wall colorless, with the forming wall-borders rose-red.

On the submerged parts of aquatic plants and dead sticks, and swimming free in fresh water.

The families consist at most of thirty-two cells; those of eight globular cells measure .00016 in. long, .000175 in. wide; tabular families of eight cells .00025 in. long, and .000175 in. wide, while those of sixteen cells are .00005 long and .00043 wide.

Probably also *Merismopedia rialacea* (Breb., Kützing, Spec. Alg. p. 472, und Tab d. phyc. v. T. 38, Fig. 7; Rabenhorst Flora Europ. Alg. II, p. 57) belongs to the Fungi. This agrees in its large size nearly with *Sarcina Reitenbachii*, yet differs on account of the color, especially as not rarely one hundred and twenty-eight cells are united in a family. Very similar, but hitherto perhaps only found in Sweden, is *Merismopedia chondroideum*, Wittr. (Wittr et Nordstedt. Algæ Exsiccata. 209).

S. hyalina, Winter.

Synonym: *Merismopedia hyalina*, Kütz.

Cells globular, almost colorless, .00006 in. in diameter; families mostly of four to twenty-four (rarely more) united cells, attaining .00060 in. in diameter.

In swamps.

Sarcina renis. Hepworth (Mic. Jour., v. 1857, p. 1, pl. 1. Fig. 2), is bright-green; it shows little connection with the species of this genus, therefore I shall only mention it.

Besides the foregoing species of *Sarcina* there are those hitherto classed as fungi on various substrata; on boiled potatoes (in little chrome yellow masses,) on cooked white of egg (bright yellow spots), also in solutions, and even in the blood of healthy and diseased human beings. Compare Cohn's Beitr. zur Biologie I, Bd. 2, p. 139.

BACTERIUM, Cohn.

Cells short-cylindrical, long-elliptical or fusiform; increasing by transverse division; having the power of moving spontaneously. The daughter cells separate soon after the division or remain attached in pairs or in a greater number forming a chain; frequently forming zoogloea; formation of spores similar to that of *Bacillus*.

[Aside from their shape the species of this genus mainly differ from those of *Micrococcus* in their power of movement. While the latter oscillate and tip and turn in the suspending fluid, but make no advance, the former freely move from place to place. Of these the motion is of every kind: rolling, spinning, turning end over end, swaying as if attacked at one end by an invisible thread, quivering without change of place, sailing steadily and stately, during like a flash, whirling, bobbing, dancing—a maze and labyrinth of movement. But these active motions are observed only when the organisms are in a rich nutritive fluid and are supplied with free oxygen. The zoogloea differ from those of *Micrococcus* in having a firmer and more abundant intercellular substance.]

B. termo, Dujard.

Synonyms: *Monas termo*, Muller (Infus. T. I, Fig. 1—non Ehrb.); (?) *Pulmella infusionum*, Ehrb. (Inf. p. 526). *Zooglœa termo*, Cohn (Nova Acta Acad. Caes. Leop. Carol. XXIV, Bd. 1, p. 123, T. 15, Fig. 9).

Exsiccata: Thümen Mycotheca Universalis, 1000.

Cells short-cylindrical, oblong; .00006 to .00008 in. long [.00008 to .00012 in. long, .00025 to .00012 in. wide, Magnin]; furnished at each end with a cilium.

In all putrescible substances, especially water in which meat is macerated.

Bacterium termo is the ferment of putrefaction; it produces putrefaction of organic sub stances, and rapidly multiplies itself as long as the substances capable of putrefaction are present, while it disappears as soon as the decomposition is ended. It is certainly obtained when a piece of meat is put into water in a vessel left open and allowed to stand in a warm place. The reproductive power is so enormous that the bacteria cells reaching the liquid through the air, or as attached to the meat, have, in a short time, so great a progeny that in twenty-four hours the water has a conspicuous milky appearance, caused by the swimming fungi. That the *Bacterium* is the cause, and not simply an accompaniment of the putrefaction, is made apparent by a simple experiment. Putrefaction begins as soon as air is allowed to freely pass to a putrescible substance, because the air always contains a number of the bacterium cells. But when the putrescible substance is heated over 50° Cent., and the air excluded, it does not ferment. It can be objected that the air, or the oxygen of the same, causes the putrefaction, but this can be readily refuted. Air filtered through cotton-wool, and thus freed from the bacterian cells, can easily be passed to a putrescible liquid which has been highly heated; in this case no putrefaction takes place.

B. lineola, Cohn.

Synonym: *Vibrio lineola*, Müller (Vermium Historia, p. 39).
Bacterium triloculare, Ehrb. (Infs., p. 79).

Cells similar to those of *Bacterium termo*, but larger; .00012 to .00015 in. long by .00006 in. wide; with two cilia at one end.

In various infusions without producing a special fermentation.

[Takes the form of zoöglœa in which the rods are motionless. Its protoplasm is often studded with dark granules. This is the only statement known to the translator concerning the *two cilia*, and his own observations have failed to confirm it.]

B. litoreum, Warming.

Cells ellipsoidal or elongated, usually rounded at the ends, .00008 to .00024 in. long, .00005 to .00J1) in. wide; colorless; motile or still, but never united in chains or zoogløa, nor in large masses.

B. fusiforme, Warming.

Cells fusiform with sharpened ends .00008 to .00020 in. long, .00J02 to .00003 in. wide.

In a loose layer on the surface of sea water.

B. navicula, Reinke and Berthold. (Die zetsetzung der Kartoffel durch Pilze, p. 21, T. VII., Fig. 10.)

Cells fusiform or elliptic, diminished at each end; rather large; motile or at rest; having within one or more opaque granules which are colored blue with iodine.

In moist-rotting potato.

B. synxanthum, Schrœter.

Synonyms: *Vibrio synxanthus,* Ehrb. (Bericht uber die Verhandl. der Berl. Akademie 1840, p. 202, No. 51. *Vibrio xanthogenus,* Tchs. (Magaz. f. d. ges. Thierheilkunde, Bd. VII, p. 194.)

Not different in form from *Bacterium termo*; .0..003 to .C0004 in. long; motions lively; single or as many as five united in a chain. Causing the so-called yellow milk.

Boiled milk becoming after a time coagulated often suddenly turns citron-yellow; the casein gradually disappears until very little remains. The milk, at first neutral becomes sour, then intensely alkaline. The filtered citron-yellow liquid becomes by evaporation amber-yellow; the resulting yellow-brown crust is insoluble in alcohol and ether, but entirely soluble in water. Alkalies do not change its color, while acids cause instant decolorization.

B. syncyanum, Schrœter.

Synonyms: *Vibrio syncyanus,* Ehrb. d. c. p. 124 und 126, *Vibrio cyanogenus,* Fehs, d. c p. 190.

Like the preceding in form.
Producing "blue milk."

The coloring matter is changed by alkalies to peach or blood red, while acids restore the original color. Ammonia, in turn, only changes the blue to a violet tint.

B. æruginosum, Schrœter.

In the so-called green (or blue) pus sometimes found in wounds, etc.

The cells themselves are also colorless in this case; they impart the verdigris green often changing into blue coloring matter to the surrounding medium.

'B. punctum, Ehrb.

Cells elongated, ovoid, colorless, often in pairs; movements slow and oscillating; length .00021 in., thickness .00007 in.

In infusions of animal substances.

B. catenula, Duj.

Cells cylindrical filiform, often three, four or five united; length
.00012 to .00016 in., thickness .00002 in.

In fetid infusions and in typhoid fever (Coze and Feltz.)

This and the preceding are taken (translated from the French) from Magnin's work on
Bacteria, but they are put down as doubtful species. They seem from the shape of the
cells to belong rather to *Bacillus*.

The four following, also copied from Magnin, are said by him to
require further study, though apparently species of *Bacterium:*

Vibrio lactic, Pasteur.

Cells almost globular, very short, a little swollen at the extremities; length, .000064
in. in a series; .00200 in. long.

Develops, according to Pasteur, in sweet liquids, in which it causes the formation
of acetic acid and in milk the coagulation of the casein. According to other researches the coagulation of the casein is influenced by a soluble (zymase), and not
an organized, ferment.

Mycoderma aceti, Pasteur.

Synonym: *Ulvina aceti.* Ktg.
Exsiccata: Thümen Mycotheca Universalis 1599.

Cells short, narrowed in the middle, often united in long chains, forming a pellicle on
the surface of liquids; length of a cell .00006 in., which is two or three times the width.

This species is thus very near the preceding; it should not be confounded with *Mycoderma vini*, which may develop in the same liquids, but which belongs to the *Saccharomycetes*.

Vibrio tartaric right, Pasteur.

Cells globular, short, .00001 in., united in chains about .00200 in. long; similar to the preceding.

Decomposes racemic acid, causing right tartaric acid to disappear, and liberating left
tartaric acid.

The acid fermentation of beer.

Cohn thinks this is due to a *Bacterium* similar to *B. termo*, but a little larger. He has
found it with oval *Saccharomycetes* in acid beer—elliptical *Bacteria* endowed with motion,
often united in pairs, rarely in fours].

BACILLUS, Cohn.

(Beitr. z. Biol. I. Bd. 2, Heft. p. 173).

Cells elongated cylindrical, almost always attached together in
straight rod-like (*stielrunden*) rows or threads (not or little interlaced); multiplying by transverse division. They form zoogloea, but
often also occur united in thick swarms without gelatinous secretion.
Propagation by spores.

The genus *Bacillus* is closely related to *Bacterium*; especially is *Bacterium lineola* with
united cells very similar to *Bacillus* rods. Yet there is this difference, that in the longer
Bacterium cells the appearance of dividing is perceptible, while in the *Bacillus* cells of
equal length it is not.

Some of the species are always motionless, some are spontaneously motile, but go into
a resting condition. The rod-like cells elongate by intercalary growth to about double
the typical length, and then divide by a transverse partition into two daughter cells, which
often separate from each other, but often also remain attached. When the products of repeated division continue joined together filaments are produced, which are zigzag or

straight, apparently jointless, but the cells become apparent by the use of coloring matters. In the formation of spores the greater part of the cell-contents collects in one place of the rod, which often swells at the point, and the protoplasmic contents becomes sharply defined from that of the rest of the cell. Later, this highly refractive, dark-appearing body (the spore) separates from the sterile part of the cell, and falls to the ground; the ripened spore continues to remain apart. These spores possess the ability to endure unfavorable influences of various kinds, without detriment to their vitality. They can remain a long time in the soil, often many years, before beginning to grow, but also have the power to germinate at once. In germination, the spore first loses its polished appearance and swells a little; the cell wall then splits around the middle of the spore. Through the opening thus made the spore protrudes by the arching of its substance, and grows into a new rod, to the base of which the old cell wall of the spore adheres, and is often thrown off only at a late period.

The determination of the different species is also here very difficult.

B. subtilis, Cohn.

Synonym: *Vibrio subtilis*, Ehrb. (Infs. p. 80, No. 91. T. V., Fig. 6).
Exsiccata: Thümen Mycotheca Universalis, 1200.

Cells cylindrical, about twice as long as thick, attaining .00024 in. long; bearing a cilium at each end. Mostly several cells, joined into apparent filaments which are motile, flexible and are furnished at each end with a cilium. The spore-forming cells are three to four times as long as thick, isolated or united in filaments. The spores are usually somewhat greater in diameter than the rods.

In different infusions and substances—very probably in the rennet stomach of living ruminants. According to Cohn, the producer of butyric fermentation, and also the active principle in the ripening of cheese.

The extraordinarily great power of resistance of the spores of Bacillus subtilis and the other species is a peculiar property. They are not killed by boiling, but made to germinate very quickly, though the duration of the boiling must of course be considered. Fifteen minutes boiling does them absolutely no harm, while most are killed after one hour and all after two hours' boiling. They are insensible to poisons and weak acids.

B. tremulus, Koch.

Very similar to the preceding but more slender and mostly shorter, always with a cilium at each end. Spores plainly thicker than the cells, often arranged in lines.

On the surface of foul vegetable infusions, forming a thick slimy layer.

B. amylobacter, Van Tieghem.

Synonym: *Closterdium butyricum*, Praz.
Exsiccata: Thümen Mycotheca Universalis, 1800.

Morphologically similar to *Bacillus subtilis*, but distinguished by the fact that at certain times it contains starch in its cell-contents as can be easily proved by the addition of iodine.

In the cells of plants having milk-sap, in foul vegetable infusions, etc.

According to Van Tieghem's first communication this species is the producer of cellulose fermentation. Later *Bacillus amylobacter* not *B. subtilis*) was pointed out by him and Praznmowski. Botan. Zeitung. 1879. No. 26 as the cause of butyric acid fermentation 'Vibrion butyrique, Pasteur). According to Praznmowski, *B. amylobacter* is distinguished especially and less intimately from *B. subtilis* in the manner of the germination of the spores. In the first species the germinal tube does not appear at the equator but at one end of the spore. But to found a new genus upon this, as proposed by Praznmowski, does not seem to me advisable.

[Treonl has held that this organism originates within the closed cells of plants by a direct transformation of the protoplasm an idea combatted by Van Tieghem. Comptes Rendus, t. 88, p. 205; t. 61, pp. 156 and 156; t. 65, p. 513].

B. ulna, Cohn.

Filaments thicker than in *Bacillus subtilis*, somewhat flexible, with dense, finely granular protoplasm. A single cell attaining .00040 in. length, .00098 in. wide. Spores oblong cylindrical. In various infusions, for example in the white of egg.

Appears to be scarcely different from *Bacillus subtilis*; intermediate forms between them have been observed.

B. anthracis, Cohn.

Exsiccata: Thümen Universalis. 1499.

Very similar to *Bacillus subtilis*, but motionless and without cilia; cells .00016 in. long and longer, very slender, mostly extended; often united in crooked filaments. Spores not, or but little thicker than the threads.

In the blood of animals which have died with splenic fever (Anthrax, Miltzbrand); the cause of splenic fever in cattle, sheep, etc., and malignant pustule in man.

Bacillus anthracis and the disease symptoms caused by the organisms are, among all pathological processes induced by Schizomycetes, the most accurately investigated. The *Bacilli* are found without exception in the blood of animals dead from splenic fever and the proof has now been sought and found that they are the *cause* of the disease. So long as only the vegetative rods were known it was difficult to gain this evidence; for these retain their vitality only a comparatively short time and blood containing only these soon loses its power of infection. The remarkable thing about splenic fever is that it often occurs very suddenly in a region, then disappears for a long time to reappear just as unexpectedly, without any transmission having been allowed. From this fact it is to be inferred that the contagion can retain its virulence a long time. The discovery of the spores of *Bacillus anthracis*, which form only in the blood of dead animals or when the blood of animals sick with splenic fever has been a long time dried, explains this power of long duration. For as the spores of *Bacillus anthracis* possess a great power of resistance to outside influences, especially dryness, they are capable of developing after many years. They are often produced within the buried bodies of animals dying with the disease, and from these they may be diffused in various ways. Then if in any manner they reach the bodies and gain entrance to the blood of cattle, etc., they germinate, reproducing the rods which multiply richly and soon begin their destructive activity.

[Recent investigations (1881-1882) have added further information upon and new interest to this species. It seems well established that *Bacillus subtilis* may, by graded cultivations, be physiologically changed, so that it is capable of developing in the blood of living animals and thus become the cause of disease; but such changes do not take place suddenly and seldom occur in nature, though the possibility of the latter may explain what has heretofore been mysterious and perplexing. But by far grea er scientific and practical interest is attached to the results of modifications through artificial cultivation of *Bacillus anthracis* itself. By cultivating the deadly organism in well aerated chicken or other broth at a certain temperature, the virulence of its physiological effects is gradually lost, but may be restored after several generations by equally feasible methods. The practical importance of this is at once seen to be very great when it is further made known that the organism modified by habit to a harmless condition constitutes a protective virus which, after inoculation, relieves the animal from danger however exposed to the *Bacillus* in its malignant state. Some account of this has been given on a preceding page, and the matter has been widely published in recent periodical literature. No greater contribution has ever been made to pathological and medicinal knowledge, and the good results already attained opens boundless anticipations of mastery over other ills that flesh is heir to. Pasteur supposed he had conclusive proof that the spores buried with dead animals retained their vitality during ten or more years, but the changes now known possible in other species renders the evidence less valuable.]

B. ruber, Frank and Cohn.

Exsiccata: Rabenhorst. Algen 2441.

Rods .00024 to .00032 in. long, scarcely .00004 in. thick, actively moving, isolated or united in twos or fours. Rods (cells) just divided sometimes shorter, only .00012 to .00016 in. long. Imparting a brick-red pigment, different from that of *Micrococcus prodigiosus*. On boiled rice.

B. erythrosporus, Cohn.

Motile, short, slender rods, forming sometimes longer filaments in which originate numerous, oblong-oval, highly polished, dirty-red spores.

On solutions of beef extract, putrid infusions of white of egg and of meat.

This species forms in part little swimming scales, in part a continuous pellicle; the filaments at length decompose into a gelatinous mass whereby the spores are liberated, which now united in little gelatinous masses sink to the bottom. The species is easily recognized by the dirty-red color of the spores.

[B. of tuberculosis.

Cells very slender, cylindrical, about .00002 in. wide, .00010 to .00012 in. long, isolated or in chains of a few articles; motionless; sometimes containing spores which, from their size, cause slight fusiform swellings of the containing cell.

That this dreaded scourge of the human family as well as of the higher animals is infections has of late been repeatedly shown, and the most careful search has been made for the organism which, from analogy, was supposed to constitute the *materies morbi*. After many failures on the part of numerous investigators, Dr. R. Koch, of Berlin, succeeded, by a special method of preparation, in discovering the minute species characterized above. In order to see the Bacillus it is first stained as follows: Smear a cover glass with the tuberculous matter 'sputa or a small portion of tubercle), dry over a lamp; float the smeared cover several hours (24) on a concentrated alcoholic solution of methylene-blue 1 part, a ten per cent. solution of potash 2 parts, distilled water 200 parts; wash, and treat with a few drops of aqueous solution of vesuvin. The Bacilli retain the blue, while the rest of the material does not. Now that we know what to look for, the organisms can be seen without staining, but they are very indistinct. The common violet stain of the accompanying material somewhat aids the Bacilli showing white.

Abundant experiments by Koch and others, have shown that these Bacilli are the true agents in the wasteful processes of the disease—the real cause of consumption in man and animals. It is also demonstrated that they do not develop in nature outside of the living body, hence that the disease is only communicated from the diseased; and, further, that the supposed hereditary peculiarities consist simply and only in the organic inability to resist infection. The children of consumptive parents may remain healthy if kept away from diseased individuals and their excretions].

[B. lepræ, Hansen.

Cells slender, elongated, .00016 in. long, .00004 in. wide; isolated or united in chains of a few articles, often arranged side by side; motionless.

In any or all tissues of the body of those afflicted with leprosy.

The investigations of Hansen, Neisser and others, have fully established the cause of this scourge of the human family in various parts of the world. The contagious character of the disease was among the earliest recognized, and has long been fully understood; but in what the contagion consists, has been entirely unknown until o ir own time. It is now added to the increasing list of known affect ations due to the injurious activities of minute parasites which we are just beginning to know and understand. *Bacillus lepra*, like the preceding, is nearly invisible without staining, but is readily seen after treatment with aniline dyes in the manner just given'.

[B. of foot-rot in sheep.

Cells cylindrical .00012 to .00016 in. long, isolated or more generally united in pairs, of which one cell is larger than the other; actively motile.

In pustules in tissues of animals affected with above named disease.

The course of the disease occupies about thirty-five days. In the vegetative stage the organisms are very active, but in liquids (broth of rabbit or mutton) from which the nutriment is nearly exhausted the larger cell of a pair produces a spore at each end and sometimes one in the middle, the smaller cell produces one spore of larger size, about .00004 in. in diameter. The spores, as in other cases, sink to the bottom as a white sediment. Upon inoculations with this material pustules are formed which reach their greatest size in about eighteen days. They never suppurate and appear to be local in effect, though the temperature of the animal rises somewhat by the fifteenth day. (Comptes Rendus, xcii, 1881, pp. 362-4.)]

LEPTOTHRIX, Kurzing (Emend).

Filaments very long and slender, unbranched, apparently not jointed, colorless, motionless, without granules, free or felted.

The fungi referred to the genus *Leptothrix* are, with reference to their specific value, very questionable; I place here the following kinds only provisionally. *Leptothrix* species very commonly occur with those of *Bacillus*. Since the genus will probably have a place among the fungi only a short time, I will not give it a new name. The greater number of the species are typical phycochromous algæ. [Treatment with iodine renders the articulations very distinct].

L. buccalis, Robin.

Filaments very long and slender .000028 to .00004 in. (seldom something more) in diameter; jointless, colorless; felted into dense white masses.

Mixed with *Micrococci* (usually also *Vibrio*, etc.,) in the white slime on the teeth, on the epithelium of the mouth and in hollow teeth. Probably the cause of caries (rotting) of the teeth.

The seat of the fungus is especially in the canals of the tooth-bone (the dentine pipe); but it seizes also upon the substance of the enamel which it gradually destroys. In the canals the fungus produces marked enlargement, later the walls themselves become penetrated by fissures and chinks and broken in pieces.

L. parasitica, Kütz.

Filaments very slender, mostly curled and crisped, obscurely jointed; loosely felted, nearly colorless; .004 to .0056 in. long, .00004 in. thick.

Parasitic on *Scytonemaceæ* and other related *Algæ*.

Leptothrix pusilla, Rabh, and *L. lanugo,* Kütz. are, perhaps, also to be accounted Fungi.

BEGGIATOA, Trevisan.

Filaments very long but thicker than those of *Leptothrix*, usually obscurely jointed, quite rigid but actively oscillating, imbedded in jelly, colorless, with numerous highly refractive granules in the protoplasm which consists of sulphur.

The genus *Beggiatoa* is easily recognized by the chalk-white slime-forming, actively moving filaments, whose joints cannot as a rule be distinguished without special processes. In order to see them it is necessary to let the filaments dry on the microscopic slide and then apply bisulphide of carbon, which gradually dissolves the granules of sulphur which in the living plant obscures the joints. The species of *Beggiatoa* live for the most part in thermal

—10

sulphur springs where they decompose, the dissolved sulphur compounds in the water and give off free sulphureted hydrogen. For this reason such water with *Beggiatoa* put into a stoppered bottle, develops an extremely intense odor of sulphureted hydrogen.

The supposed species of *Beggiatoa* are of very uncertain value; they are distinguished by very little else than the diameter of the threads.

B. alba, Trev.

Synonyms: *Beggiatoa punctata*, Trev. (Flora Enganea. p. 56); *Oscillaria alba*, Vauch (Conferv. p. 198, T. XV. Fig. 11); *Hygrocrocis Vandelli*. Menegh. (Kutzing's Algæ exsc. No. 16—Tab. phycol. I, T. XXXVIII, Fig. 3).

Exsiccata: Wartmon und Schenk, Schweiz. Kryptog. 689.

Filaments without evident joints, forming dirty or chalky-white slimy masses; .00012 to .00014 in. thick.

In sulphur springs and swamps.

Var. marina, Cohn.

Filaments densely filled with blackish granules, only .00008 thick.

In a salt-water aquarium, forming a snow-white, then slimy membrane on dead animals and algæ.

B. nivea, Rabh.

Synonym: *Leptonema niveum*, Rabh. (Alg. Decad. 653).

Filaments very slender, obscurely jointed; .00004 to .00006 in. thick (according to Rabenhorst) forming flocks of snow-white color.

In sulphur springs.

In Wartmann and Schenk's Swiss Cryptogams, 689, this species is given under the name of *Symphyothrix nivea*, Brugger. Both the above names are cited, *pari passu*, as synonyms. I take the following from their description: "Filaments not polished, without joints, also without movement, only .00002 to .000052 thick, parallel and variously entwined, united in pencil-like tufts, strings and bundles of very unequal thickness, which are enveloped in a common, homogeneous slime-mass."

B. leptomitiformis, Trev.

Synonym: *Oscillaria 'leptomitiformis*, Menegh. (Ragazz. Nuovo ricerch. fisico-chim. p. 122.—Kutzing, Tab. phycol. I, T. XXXVIII, Fig. 1.)

Exsiccata: Rabenhorst's Algen, 1813.

Filaments very slender, obscurely jointed, .00007 to .00010 in. thick; forming a thin chalk-white slimy layer.

In sulphur springs.

B. archnoidea, Rabh.

Synonyms: *Oscillaria archnoidea*, Ag, (Regensb. "Flora," 1827, p. 634, No. 38.) *Oscillaria versatilis*. Kutz. Phycol. gener. p. 180).

Filaments rather thick, evidently jointed, with rounded, slightly curved ends; movements active. Joints as long, or half as long as thick. Filaments .00020 to .00026 in. thick, forming a very thin, cob-webby, chalk-white, slimy pellicle.

In sulphur springs and swamps.

55

B. pellucida, Cohn, (Hedwigia 1865, p. 82, T. I, Fig. 2.)

Filaments .00020 in. thick, motile, evidently articulated, with rounded ends; joints scarcely as long as thick; pellucid, containing but few granules.

In a salt-water aquarium.

B. mirabilis, Cohn, (l. c. p. 81, T. I, Fig. 1.)

Filaments thick, variously crooked and curled, with rounded ends, evidently articulated; attaining .00064 in. thick; joints half as long as thick, filled with numerous rather large granules. Threads twisted and woven through each other, forming a snow-white, slimy web.

With the preceding.

DOUBTFUL SPECIES.

B. tigrina, Rabh. (Flora Europ. Alg. II, p. 95.)

Synonym: *Oscillaria tigrina*, Römer (Die Algen, Deutchlands p. 58.)

Filaments rather thick, oscillating, evidently jointed, with slight and obtuse curves, now and then suddenly reduced in size, with rounded ends; pellucid, .00014 to .00018 in. thick; forming a thin white layer.

In swamps and on wood under water.

B. minima, Warm.

Very small, flexible and actively moving; the longest .0016 in..00007 to .00008 thick; jointed; distinguished by the order of the delicate stripes. Each joint about half as long as wide. Without granules.

In sea-water.

CLADOTHRIX, Cohn.

(Streptothrix, Cohn, Beitr. z. Biol. Bd. I, Heft 3, p. 204.)

Filaments leptothrix-like, very slender, colorless, without joints, straight or slightly undulating, or irregularly spirally wound, with apparent branches.

I am unable to find a satisfactory difference between the genera *Cladothrix* and *Streptothrix*. Both are very doubtful genera. Compare Cienkowski's "Zur Morphologie der Bacterien" (Memoires de l' Acad. imp. d. Sciences de St. Petersbourg. VII Sér. Tome XXV, No. 2, p. 11.)

C. dichotoma, Cohn.

Filaments repeatedly dichotomously branched, straight or slightly bent, .000012 thick, forming little webs (Räschen) .02 in. and more in diameter.

In foul water, sometimes floating on the surface, sometimes attached to algæ.

The branching in this case, as with *Cladothrix Forsteri*, is only apparent. The filament split themselves into two halfs, which independently elongate and so grow side by side; in this way the separated pieces are crowded to one side and appear as branches.

C. Forsteri, Winter.

Synonym: *Streptothrix Forsteri,* Cohn. Beltr. z. Biol. I, 3 Heft., p. 186 und 204.

Filaments straight or curved, irregularly spirally twisted, sparsely and irregularly branched, occurring in pieces of various lengths.

In the lachrymal ducts of human eyes, forming greasy or crumbling yellowish-white or blackish concretions, .125 to .25 in. long, .083 in. thick.

MYCONOSTOC, Cohn.

(l. c. p. 183 und 204.)

Filaments very slender, colorless, not jointed, but upon drying separating into short cylindrical articulations, variously curved and entwined, imbedded in jelly, forming globules from .00040 to .00068 in. in diameter.

Multiplying by the infolding and the division of the globules of jelly in two parts.

M. gregarium, Cohn.

Gelatinous globules, floating on the surface of foul waters, single or aggregated in little slimy drops, with the circumference sharply defined.

On water in which there are decomposing algæ.

SPIROCHÆTA, Ehrb.

(Abhaudl. d. Berlin. Acad. 1833, p. 313.)

Cells united in long slender threads, mostly showing narrow spiral windings. The filaments have the liveliest movements, and clearly propel themselves forward and backward, but are also able to bend in various ways. Forming no zooglœa, but often felted in dense clusters. Differs from *Spirillum* by the long, narrowly-wound, flexible threads.

S. plicatilis, Ehrb.

Synonym: *Spirillum plicatile,* Duj. (Inf. p. 225, T. I, Fig. 10).

Spirulina plicatilis, Cohn. (Nova Acta Acad, Caes. Leopold; Carroll, XXIV, I, p. 125, T. XV, Fig. 10, 11.

Filaments very short and slender, with numerous narrow spiral turns; jointed, obtuse at the ends, .00410 to .00900 in. long (according to Rabenhorst); diameter, .00009 in. (according to Ehrenberg). In swamp water, among algæ.

This species differs, according to Koch, from the others, by the double spiral formed by the filaments. Yet threads wound in a continuous spiral are very common.

S. Obermeieri, Cohn.

Very similar to *Spirochæta plicatilis* in form, only differing by the filaments being sharply pointed at both ends.

In the blood of the sick with recurrent fever, and apparently the cause of the sickness.

The filaments of *Spirochæta Obermeieri* are either extended and regularly spirally wound, or they are bent, so that the spirals appear irregular, especially in the parts most crooked, rapidly moving in various ways. This species is found in the blood of persons having recurrent fever, and really only during the returning access (onset) of the fever or a short time thereafter. They disappear in the periods between the paroxsms of the fever. [The former investigations have been confirmed by extensive observations in India, where relapsing fever is now very common. See volume on "Spirillum Fever," by Dr. Vandyke Carter, London, 1882.]

S. Cohnii, Winter.

Very similar to both the foregoing species, but always shorter and usually also more slender, like *Spirochæta Obermeieri* sharply pointed at both ends.

In mucus on the teeth; discovered by Cohn, figured by Koch (Beitr. z. Biol. II, Bd. 3, Heft., T. XIV, Fig. 8).

S. gigantea, Warming.

Filaments cylindrical, obtuse at both ends, about .00012 in. thick, with numerous spiral turns of a diameter of .00028 to .00036 in. and length of each .001 in.; flexible; the articulations are not apparent, but the threads sometimes separate into joints. The longest have sixteen spiral turns; cilia have not been found.

In sea water.

SPIROMONAS, Perty.

(Zur Kenntniss der Kleinsten Lebensformen. p. 171.)

Cells leaf-like (flat), compressed, twisted around an ideal longitudinal axis; multiplying by transverse division.

S. volubilis, Perty.

Colorless, pellucid, polished, without in any part special differentiation; movements quite rapid, turning upon the axis upon which the leaf-like body is wound. Body often very little twisted, never forming more than one spiral turn; length .00060 to .00070 in.

In stagnant swamp water and foul infusions.

S. Cohnii, Warming.

Cells flattened, but sometimes scarcely angular, sharply pointed at both ends and always furnished with a cilium, having one and a fourth (rarely more) spiral turns; these six to nine times as high as their diameter; height .00036 to .00080 in., diameter .00005 to .00014 in. Thickness of the cells .00005 to .00016 in.; colorless, often with one or two longitudinal stripes.

In ill-scented, very strongly fermenting water.

SPIRILLUM, Ehrb.

(Abhand. d. Berl. Akad. 1830, p. 38).

Vibrio. Cohn. (Beitr. z Biol. I, Bd. 2, Hoft. p. 178.)
Ophidomonas. Ehrb. (Inís. p. 43.)

Cells cylindrical or somewhat compressed, with a single arch-form curve or spirally wound; rigid; furnished at each end with a cilium (not certainly observed in all the species); multiplying by transverse division, the parts soon separating from each other. The formation of zoöglœa and of spores as in the species of *Bacillus* sometimes occurs.

I unite with *Spirillum* the genera *Vibrio*, Cohn, *Ophidomonas*, Ehrb. The genus *Vibrio* does not indeed permit of sharp definition since their cilia have been found. Cohn himself has united *Ophidomas* with *Spirillum*. Warming also shows that all three genera are the same. Although the name *Vibrio* has the priority, I have chosen *Spirillum* because with the first, aside from its being non-botanical, misuse has been practiced, so that it is better to drop it altogether.

S. rugula, Winter.

Synonyms: *Vibrio rugula,* Muller (Infs. p. 44, T. VI, Fig. 2.) *Melanella flexuosa,* Bory (Encycl. method. 1824.)

Cells .00024 to .00064 in. long, .00002 to .00010 in. thick; either only one curve or one flattened spiral turn, bearing a cilium at each end, actively rotating around their long axis; the cells often felted into dense swarms; height of a spiral mostly .00024 to .00040 in., diameter .00004 to .00008 in.; globular spores always formed at the ends of the cells.

In swamp water and various infusions; also in the slimy material on the teeth, &c.

According to Warming some specimens attain a height of single spiral of .00050 to .00080 in., and diameter of .00010 to .00020 in.

S. serpens, Winter.

Synonym: *Vibrio serpens,* Muller (Infs. T. VI, Fig. 7 and 8.)

Cells half as thick as the preceding species, .00045 to .00112 in. long (according to Rabenhorst) .00003 to .000045 in. thick, with more usually three to four spiral turns; often joined in long chains; furnished with a cilium at each end; also often collected in swarms; height of a single spiral .00030 to .00050 in.; diameter, .00005 to .00012 in.

In various infusions.

Rabenhorst's measurement of the length, .00002 to .00112 in., probably applies to the whole filament consisting of several cells. According to Warming the height of a single spiral sometimes attains .00088 in.

S. tenue. Ehrb. (Infs. p. 84, T. V, Fig. 11).

Cells very slender, .00016 to .00060 in. long, .00010 in. thick (according to Ehrenberg), with at least one and a half, usually two,

three, four or five spiral turns; height of a single turn of the spiral, .00006 to .00016 in.; diameter, from one-half to the same; movements very active, but also without motion; collected in dense swarms or masses, or forming zooglœa.
In various infusions.

According to Warming only .00004 in. thick, and the distance of the turns of the spiral sometimes .00032 to .00040 in.; their diameter only one-eighth to one-tenth this measurement. A confusion appears to prevail in respect to *Spirillum tenue* and *Sp. undula*.

S. undula, Ehrb.

Synonyms: *Vibrio undula*, Müller (Vermium historia, p. 43). *Vibrio prolifer*, Ehrb. (Inf. p. 81, T. V, Fig. 8.)

Cells .00032 to .00018 in. long, .000044 to .000056 in. thick (Rabenhorst; spiral wider than the preceding; turns .00016 to .00020 in. distant; each cell usually having only one-half or one, rarely two or three spiral turns; furnished at each end with a cilium; motions very active; sometimes also forming zooglœa.
In swamp water and in various infusions.

Ehrenberg gave for *Spirillum tenue* a thickness of 1-1000 of a Prussian line, and for *Sp. undula* only 1-1680 of a line; he also said in the description: *"Sp. fibris valde tortuosis brevibus, validioribus."*
According to Warming, *Spirillum tenue* is more variable than has been hitherto supposed. The turns of the spiral are often very long, so that the cell appears almost straight, therefore the distance between them varies from .00012 to .00042 in., with a diameter from one-tenth to three-fourths of this measurement; the thickness of the cell is from .000024 to .00005.

Var. litorale, Warming.

Attains .00012 in. thick, length of one turn of the spiral form .00020 to .00040 in., and diameter from one-sixth to one-fourth as much.
On the coast of the Baltic sea.

S. volutans, Ehrb.

Synonyms: *Vibrio spirillum*, Müller (Inf. p. 49, T. VI, Fig. 9). *Melanella spirillum*, Bory (Encycl. method.)

Cells somewhat tapering at the ends, gradually rounded, .00100 to .00120 in. long, .00006 to .00008 in. thick; each cell with two and a half to three and a half spiral turns, each of which is .00036 to .00050 in. high, .00026 in. in diameter; furnished with a cilium at each end.
In various infusions as well as in swamp water among algæ.
According to Warming, the spiral is often elongated so that the cell appears almost straight, the diameter then becoming only .00006 to .00016 in.

Var. robustum, Warming.

Thickness .00008 to .00018 in.; height of spiral .00040 to .00080 in.; diameter .00004 to .00012 in.; mostly one and a half turns; sometimes two cilia at one end.
In sea water.

S. sanguineum, Cohn.

Synonym: *Ophidomas sanguinea*, Ehrb (Monatsber. d. Berl. Akad. 1840, p. 201).

Cells cylindrical, only rarely tapering at the end, .00012 in. and more thick, of various lengths, with mostly two, seldom only one-half or two and a half spiral turns; distance of the latter .00036 to .00048 in.; diameter two-thirds as great; furnished at each end with a cilium. Cell-contents colored by numerous reddish granules with many granules of sulphur.

In foul brackish water.

According to Warming the longest specimens reach a length of .00280 in., the distance of the turns of the spiral .00060 to .00150 in., diameter one-half to two-thirds, or with the smallest one-fourteenth to one-seventh as much.

S. violaceum, Warming.

Cells either crescent-form (without a complete spiral winding) or with one or one and a fourth spiral turns; abruptly rounded at the ends and furnished with a cilium. Cell-contents violet, containing little granules of sulphur. Distance of the turns of the spiral .00030 to .00040 in., diameter .00004 to .00006 in., thickness of the cell .00012 to .00016 in.

In brackish water.

S. Rosenbergii, Warming.

With one or one and a half turns of the spiral, cells .00016 to .00050 long, .00006 to .00010 in. thick; colorless, but with numerous, highly refractive, sulphur granules. Distance of a turn of the spiral .00024 to .00030 in., diameter very different, highest half as much. Active and moving in various ways, but without cilia as it appears.

In brackish water.

S. attenuatum, Warming.

Cells tapering to the ends, usually with three spiral turns, the middle one is .00044 in. high and .00024 in. in diameter, the end ones .00040 in. high, and .00080 in. in diameter; thickness of the cell .00005 to .00008 in.

In sea water.

S. jenense, Winter.

Synonym: *Ophidomonas jenensis,* Ehrb (Infs. p. 44.

Cells obtuse at both ends, furnished with a cilium, olive-brown, .00160 in. long, .00132 in. thick, with one-half to two and a half spiral turns.

Whether this is really a distinct species is hard to say so long as it is not again found in the original locality. Possibly it is identical with *Spirillum volutans.*

APPENDIX.

We connect with the *Schizomycetes* some genera which are united with them on the part of others without question; but which show so many peculiarities that I may provisionally separate them.

SPHÆROTILUS, Kutzing.

(Linnæa VIII, 1833, p. 385, T. IX.)

Cells roundish, angular, or oblong, rounded on the angles, in the greater number united end to end in a colorless, slimy sheath, into long filaments which form cue-like, interlaced and entangled floating flocks. Multiplying by isolating vegetative cells which produce, through continued division, new filaments; propagation by spores, which form within the vegetative cells.

S. natans, Kützing, (l.c.; also 51 Jahresb. d. Schles. Ges. f. vaterl. Cultur. 1876, p. 133.)

Flocks in the vegetative stage, in the old parts yellowish-brown, in the younger colorless, much branched, very slippery. In those producing spores, a part milk-white and a part colored red. Cells .00016 to .00036 in. long, .00012 in. thick.

The flocks consist of an enormous mass of long, variously collected threads, which are formed of cells in rows, surrounded by a slimy, refractive sheath. These threads often form shrubby, branchy structures, which become attached to aquatic plants, or float in a thin stratum on the water. In the thread-cells forming spores the protoplasm separates in numerous small, highly refractive portions, which become the globular spores colored; when ripe, red; later, brown. These become free when the membrane of the mother cell dissolves. They germinate very soon and grow into threads, which are either isolate or attached to the mother or other threads. These daughter threads, formed from the germinating spore, are at first undivided, and only later become the typical row of cells. Sometimes the spores develop into threads inside the mother cell.

CRENOTHRIX, Cohn.

(Beitr z. Biol, I, Bd. 12, Heft, p. 130).

Filaments cylindrical, slightly club-form, thickened above, jointed, furnished with a sheath; multiplying by the escape of the joint-like cells from the sheath and their growth into filaments. Propagation by spores, which are formed within the sheath by the further division of the cells. The spores either grow directly into threads, or form, by continued division, gelatinous colonies of round cells, which afterward produce threads.

C. Kuhniana, Zopf.

Synonyms: *Leptothrix Kuhniana*, Rabh. (Algen Sachsens, No. 284;) *Hypheothrix Kuhniana*, Rabh. (Flora Europ. Alg. II, p. 88); *Crenothrix polyspora*, Cohn, Beitr. z. Biol. I, Bd. 12, Heft, p. 131); ? *Palmellina flocculosa*, Radlkofer. (Zeitschrift f. Biol. Bd. I).
Exsiccata: Rabenhorst's Algen, 284.

Filaments in whitish or brownish webs (Räschen), .00003 to .00020 in. thick, widened near the ends to .00024 to .00036 in.; joints of very different lengths. Spores .00004 to .00024 in. in diameter.
In springs and drainage tile, etc.

Often a very troublesome fungus since it pollutes the water and stops up small pipes. The cylindrical filaments slightly thickened toward the end are evidently articulated; the joints after a while separate from each other, but are then inclosed by a sheath, which, originally colorless, becomes yellow or yellowish brown by imbibing iron. The sheath, at first closed, is finally burst by the continued division of the joints and these escape. Each joint can develop into a new filament. But in other cases the thread remains enclosed in the sheath; its joints are divided by numerous close cross-partitions into thin disks, which then by vertical divisions separate into little globular cells; these may be considered as the spores of the fungus. They often develop inside the sheath nto now filaments which grow through the swelling gelatinous sheath; or they leave the sheath in order to further develop outside. They either grow in filaments or form by repeated bi-partition little colonies of rounded cells, which are held together by the membranes, now become gelatinous. These colonies are assigned to the *Palmella* (perhaps *Palmellina flocculosa*, Rad.); each of their cells can again form a filament.

SACCHAROMYCETES.

Saccharomycetes, or yeast fungi, are one-celled plants which multiply by budding and propagate themselves by spores produced within the cells. They live isolated or joined in sprouting chains (sprossverbänden), chiefly in liquids containing sugar in which they induce alcoholic fermentation.

In most *Saccharomycetes* the cells are globular, oval or elliptical, only rarely do they elongate into cylindrical tubes, which become jointed by transverse partitions, and may then be considered the earliest imitation of hyphæ or mycelium formation. For the purpose of multiplication the cell pushes out a little rounded protuberance (Austülpung) which becomes filled with a part of the contents of the mother cell, whose form and size it gradually acquires, and is cut off by a partition wall. Both cells can, in like manner, produce daughter cells, which frequently remain attached for a time and after their separation continue to vegetate independently.

A damp solid substance is especially favorable for the formation of spores. Typically the whole cell contents divides into two to four rounded portions, or contracts into a single globular body. Each of these surrounds itself with a cell-wall, and thus becomes a spore which can bud like the vegetative cells.

To the yeast fungi (in the narrower sense) belongs the ability to decompose the sugar of a solution, for example of wine must, into alcohol and carbonic acid, that is, to set up alcoholic fermentation. The carbonic acid escapes in rapid streams, while the alcohol, as well as some subordinate elements of sugar, e. g. succinic acid, remains behind. The fermentation proceeds with special energy with a small supply of air; but by long continued exclusion of air the yeast cells perish.

The validity of the *Saccharomycetes* species from a botanical standpoint, is similar to that of the *Schizomycetes*. As with the latter, it is also necessary here to make a limitation to the leading species, and to leave out of consideration only those species established by reliable investigators. Even then there remains much doubt, for the majority of accepted species at present are probably only different forms of one and the same kind, which have become differentiated under changed conditions of growth.

SACCHAROMYCES, MEYEN.

One-celled fungi with vegetative multiplication by budding; propagation by spores which (usually) form by the division of the contents of the mother cell.

[This is the only genus, hence has the general characteristics of the group. The relation to the *Schizomycetes* is certainly quite close and apparently nearer than usually supposed by excellent authorities. The so-called budding is, after all, only a peculiar mode of self-division by elongation and the formation of transverse partitions, and the production of spores is entirely similar in the one to the other, while the physiological processes and effects are not more distinct than the existing difference in these respects between true species of *Schizomycetes*. For these reasons, as well as the fact that the two kinds of organisms are very commonly associated in nature, I have appended this account of the *Saccharomyces* without intending to imply that the species belong among the *bacteria*.]

S. cerevisiæ, Meyen.

Synonyms: *Torula cerevisiæ*, Turpin (Compt. Rend. VIII, 1839, p. 379); *Cryptococcus fermentum*, Ktz. (Species Algarum, p. 146); *Hormiscium cerevisiæ*, Bail, (Flora, 1857, p. 417.)

Exsiccata: Rabenhorst's Algen, 121; Fungi Europ., 1999; Thümen Mycotheca Universalis, 800; Kryptogamen Badens, 141.

Cells mostly globular or oval, .00032 to .00036 in. long; isolated or joined in little colonies; spore-forming cells isolated, .00044 to .00058 in. long; spores usually three or four in a mother cell, .00016 to .00020 in. in diameter.

In beer, in both the surface and bottom fermentation.

This peculiar beer yeast is found in the various kinds of beer, in both kinds of fermentation. It is cultivated in quantity and furnishes then the so-called compressed yeast—a mass consisting of yeast cells and water.

S. ellipsoideus, Reess, (Bot. Unters. üb. d. Alkoholgährungpilze, p. 82.)

Exsiccata: Rabenhorst's Fungi Europ. 2000.

Cells elliptical, usually .00024 in. long; isolated, or united in little branched colonies. Spore-forming cells mostly isolated; spores in the mother cell two to four, .00012 to .00014 in. in diameter.

In wine must, spontaneously fermenting.

S. conglomeratus, Reess, (l. c. p. 82).

Cells almost globular, .00020 to .00024 in. in diameter, united in skeins which consist of numerous budding cells from one or a few mother cells; spore-forming cells often to one or two vegetative cells united; spores two to four in a mother cell.

In wine must at the beginning of the fermentation and on decaying grapes.

S. exiguus, Reess, (l. c. p. 83).

Cells conical or top-form .00020 in. long, .00010 in. wide, united in little branched colonies; spore forming cells isolated with always two to three spores in a row.

Among the yeast of the secondary fermentation of beer.

S. Pastorianus, Reess. (l. c. p. 83.)

Exsiccata: Thümen Fungi Austriaca 1099 (var. Rubi-Idal,) und 1199 (var. Ribis).

Cells roundish-oval or elongate-clavate, of various dimensions; colonies branched, consisting of primarily club-shaped joints .00072 to .00088 in. long, which form secondary roundish or oval, angular cells .00020 to .00024 in. long. Spore-forming cells roundish or oval; spores two to four .00008 in. in diameter.

In the yeast of the secondary fermentation of wine, ciders and self-fermenting beer.

S. apiculatus, Reess. (l. c. p. 84.)

Exsiccata: Thümen Fungi Austriaca, 263.

Cells lemon-shaped, with a little short point at each end; .00024 to .00032 in. long, .00008 to .00012 in. wide, sometimes a little longer; daughter cells only from the ends of the mother cell, usually soon isolated, rarely joined in little scarcely branched colonies. Spores not known.

In the principal fermentation of wine and other spontaneous fermentations.

S. Sphæricus. Saccardo (Michelia I, p. 89, et Fungi Ital. Autogr. del. No. 76.)

Exsiccata: Thümen Mycotheca Universalis, 900.

Cells of different forms; the basal one (of a colony) oblong or cylindrical, .00040 to .00060 in. long, .00020 in. wide; the rest globular, .00020 to .00024 in. in diameter, united in crooked, branched, often skein-like families; spore formation not known.

In the fermenting juice of *Lycopersicum esculentum* (Tomato.)

S. glutinus, Cohn (l. c. p. 187.)

Synonym: *Cryptococcus glutinus,* Fresenius. (Beitr. z. Mycol. 2 Heft. p 77.)

Cells globular, oval, oblong, elliptical or short cylindrical, .00020 to .00044 in. long, .00010 wide, isolated, or two, rarely more, united; cell wall and contents in a fresh condition, colorless, after drying and again moistened a slightly reddish nucleus in the middle; spore formation unknown.

On starch paste, slices of potato, etc., forming rose-red slimy spots which at the beginning have a diameter of .02 to .04 in., but gradually spreading and uniting they cover a surface of more than .4 in. square. The coloring matter is not changed by acids or alkalies. . ʀ '

S. Mycoderma, Reess (l. c. p. 89).

Synonyms: *Mycoderma cerevisiæ* and *M. vini,* Demaz. (Ann. Scienc. Natur. I Serie. Tome x. p. 59 et 66. *Hormiscium vini* and *cerevisiæ,* Bonord. (Handbuch p. 33, T. I, fig. i und 2.

Exsiccata: Thümen Fungi Austr. 1259, 1300.

Cells oval, elliptical or cylindrical, .00024 to .00028 in. long, .00008 to .00012 in. wide, united in richly branched colonies. Frequently the cells are elongated, mycelium-like; spore-forming cells reaching a length of .00030 in.; spores one to four in each mother cell.

On fermented liquids, sauer-kraut, juices of fruits, etc. On wine and beer, forming the so-called mold.

THE BACTERIA.

An Illustrated Monograph by T. J. BURRILL, PH. D., Professor of Botany, Illinois University.

CONTENTS.

Introduction.

PART I. -Nature and Organization of Bacteria.

(1) Existence, (2) Color, Shape, Size, (3) Movement, (4) Structure, (5) Reproduction and Development, (6) Vitality and Endurance, (7) Nutrition, (8) Origin, (9) Place in Nature.

PART II.—Effects of Bacteria.

(1) Fermentations and Putrefactions, (2) Diseases, (3) Benefits.

PART III.—Classification of Bacteria.

Sixteen Genera (with artificial key) and over one hundred Species systematically arranged and described.

EXTRACTS.

"It is the object of this paper to present, in language freed as far as possible from technical terms, the principal and most interesting facts now known about these silent-working denizens of the earth, the air, and the water."

"We swallow them with our food, and at least some kinds sometimes retain their activity in the stomach and intestinal tube. It now seems certain that the latter is always inhabited by special kinds which have to do with the activities there in operation. In health the blood is usually quite free from them, but in certain diseases this too, as it rapidly courses through the arteries and veins, sweeps along in the current myriads of the minute but living and developing, ever active things, inappropriately called "germs.""

"There is now in certain cases just as good evidence that bacteria cause diseases as there is that hawks destroy chickens, and the evidence is as inductively rigid in the one case as in the other."

Octavo. 65 pages, paper, 50 cents. Ten or more copies by express, 35 cents each. For sale by the author, Champaign, Illinois.

This and the following species reach in their development the highest rank in the *Saccharomycetes*. The cells often form especially in aqueous solutions elongated tubes, which become articulated by the growth of cross partitions and from these separate into single cells. The latter bud on their part in a similar manner.

While the proper yeast fungus vegetates submerged, in the upper strata of liquids and here sets up very active alcoholic fermentation, the mold fungus grows on the surface without exciting fermentation. Artificially forced to grow submerged there is a small quantity of alcohol formed, but the fungus soon perishes.

Although the growth of the mold-layer goes hand in hand with the souring of wine and beer, yet this *Saccharomyces* is not the cause of the latter phenomenon. Several other fungi whose systematic position is not certain, produce this vinegar out of the alcohol of wine, etc. According to some it is a *Vibrio (spirillum)* species which excites this decomposition.

S. Albicans, Reess (Sitzungsber der physic. Med. Soc. Erlangen. 9 Juli, 1877).

Synonym: Oidium albicans, Robin (Hist. Nat. d. Veget. Paras. p. 488, Pl. 1, Fig. 3 to 7).

Cells in part globular, in part oval, or elongated to cylindrical, .00014 to .00020 in. wide, the globular ones .00016 in. in diameter, the cylindrical ones ten to twenty times as long as thick. Budding colonies usually consisting of rows of cylindrical cells, from the ends of which rows of oval or globular cells are produced by budding. Spores single, formed in roundish joints.

On the mucous membrane of the mouth, especially of nursing infants, producing the disease known as Thrush ("Soor"). Also in animals.

This fungus appears in the form of less or greater grayish-white masses which, however, do not consist entirely of Saccharomyces, but also contain *Schizomycetes* and the mycelium of mold fungi. When cultivated the fungus forms abundant long-jointed, richly branched threads; at the upper end of each joint is usually found a crown or tuft of short cells which have an oval or globular form and these bud again in their turn. In other cases all the cells of a colony remain short and take the globular form. The fungus excites alcoholic fermentation only in a slight degree. According to Grawitz (Virchow's Archiv. f. pathol. Anat. und Physiol. 70 Bd. p. 557), *Saccharomyces albicans* is identical with *S. mycoderma*.

S. guttulatus, Winter. (Doubtful species.)

Synonym: *Cryptococcus guttulatus*, Robin (l. c. p. 327, Pl. IV, Fig. 2.)

Cells elliptical or elongated oval, .00060 to .00096 in. long, .00020 to .00032 in. wide; brown, opaque, with two to four colorless vacuoles, isolated or two to five united. Spore-formation unknown.

In the œsophagus and intestines of mammals, birds and reptiles.

www.ingramcontent.com/pod-product-compliance
Lightning Source LLC
Chambersburg PA
CBHW022007190326
41519CB00010B/1417